ECONOMIC AND SOCIAL COMMISSION FOR ASIA AND THE PACIFIC

Space Technology Application Capabilities, Facilities and Activities in the ESCAP Region:

A REGIONAL INVENTORY
1997

UNITED NATIONS

New York, 1998

ST/ESCAP/1868

UNITED NATIONS PUBLICATION
Sales No. E.99.II.F.45
Copyright © United Nations 1999
ISBN: 92-1-119912-3

FOREWORD

This is the first inventory on space technology capabilities, facilities and activities in the Asian and Pacific region to be published under the aegis of the Regional Space Applications Programme for Sustainable Development (RESAP), launched at the Ministerial Conference on Space Applications for Development, held at Beijing in 1994. It is part of the information dissemination and exchange efforts of ESCAP intended to foster closer regional collaboration in space technology applications, in line with the recommendations of the Conference. The publication of this inventory is part of the activities of ESCAP for 1997, as endorsed by the Commission.

The inventory provides comprehensive information on a wide range of aspects concerning space technology applications in the context of sustainable development in the region. While primarily consisting of inputs received from individual countries in response to a set of question- naires specifically designed for the purpose, it also provides overviews of space activities at regional and national levels, based on information extracted from a variety of sources, such as ESCAP publications, country reports and papers presented at meetings, seminars and conferences, national reports submitted to various forums, journals, yearbooks, newsletters and brochures. It resembles a survey rather than a conventional inventory that simply lists facilities and equipment.

It is believed that the inventory will provide a useful reference tool not only for the space community but also for other scientists and technologists, researchers, environmentalists, resource managers, planners, decision makers and the general reader. The use of space technology in support of environmentally sound and sustainable development is increasing with the passage of time as this concept gains greater acceptance worldwide, especially after the adoption of Agenda 21 at the United Nations Conference on Environment and Development at Rio de Janeiro in 1992. It is expected that the next century will witness a greatly expanded role for space technology applications in sustainable development, environment management, natural resource surveying and accounting, disaster monitoring and related activities, as space applications become integrated with sustainable development planning. The future seems to present great challenges and opportunities. It is hoped that the region will collectively take the right decisions in meeting the challenges and in making optimum use of the opportunities that will unfold.

We would like to acknowledge the substantial support provided by RESAP national focal points, national contact points and other respondents in various organizations and countries across the region in collecting and consolidating the responses to the questionnaires for this inventory and forwarding them to the secretariat. Without their cooperation and help, it would not have been possible to bring out this publication.

Adrianus Mooy
Executive Secretary, ESCAP

CONTENTS

Australia
Azerbaijan

A

Bangladesh

B

China

C

Fiji

F

India
Indonesia
Islamic Republic of Iran

I

Japan

J

Malaysia
Mongolia
Myanmar

M

Nepal
New Zealand

N

Pakistan
Philippines

P

Republic of Korea
Russian Federation

R

Singapore
Sri Lanka

S

Thailand

T

Vanuatu
Viet Nam

V

CONTENTS *(continued)*

CONTENTS *(continued)*

Page

CONTENTS *(continued)*

ANNEXES

LIST OF TABLES AND FIGURES

LIST OF TABLES AND FIGURES *(continued)*

ABBREVIATIONS AND ACRONYMS

ACRES	Australian Centre for Remote Sensing
ADEOS	Advanced Earth Observing Satellite (Japan)
AIT	Asian Institute of Technology
ANASA	Azerbaijan National Aerospace Agency
APT	Automatic picture transmission
ASEAN	Association of South-East Asian Nations
COSPAR	Committee on Space Research
COSSA	CSIRO Office of Space Science Applications
CSIRO	Commonwealth Scientific and Industrial Research Organization (Australia)
DEPANRI	National Council for Aeronautics and Space (Indonesia)
DOST	Department of Science and Technology (Philippines)
ECO	Economic Cooperation Organization
ERS	Earth Resources Satellite (ESA)
ESA	European Space Agency
ESCAP	Economic and Social Commission for Asia and the Pacific
ESCWA	Economic and Social Commission for Western Asia
GIS	Geographic information systems
GMS	Geostationary Meteorological Satellite (Japan)
GPS	Global Positioning System
HRPT	High-resolution Picture Transmission
ICSU	International Council of Scientific Unions
INSAT	Indian National Satellite
IRIMO	Islamic Republic of Iran Meteorological Organization
IRS	Indian Remote Sensing satellite
IRSC	Iranian Remote Sensing Centre
ISAS	Institute of Space and Astronautical Science (Japan)
ISRO	Indian Space Research Organization
ITU	International Telecommunication Union
JERS	Japan Earth Resources Satellite
JICA	Japan International Cooperation Agency
JPL	Jet Propulsion Laboratory (at Pasadena, California, United States)
LAPAN	National Institute of Aeronautics and Space (Indonesia)
LIS	Land information system
MACRES	Malaysian Centre for Remote Sensing
MITI	Ministry of International Trade and Industry (Japan)
MOS	Marine Observation Satellite (Japan)
MSS	Multispectral Scanner (on board Landsat)
NAMRIA	National Mapping and Resource Information Authority (Philippines)
NASA	National Aeronautics and Space Administration (United States)
NASDA	National Space Development Agency (Japan)
NOAA	National Oceanic and Atmospheric Administration (United States)
NRCT	National Research Council of Thailand
NRSA	National Remote Sensing Agency (India)
PAGASA	Philippine Atmospheric, Geophysical and Astronomical Services Administration
RESACENT	Remote Sensing Applications Centre (SUPARCO, Pakistan)
RESAP	Regional Space Applications Programme for Sustainable Development
ROK	Republic of Korea
RONAST	Royal Nepal Academy of Science and Technology
RRSP	Regional Remote Sensing Programme

RS	Remote sensing
RSA	Russian Space Agency
SAR	Synthetic aperture radar
SPARRSO	Space Research and Remote Sensing Organization (Bangladesh)
SUPARCO	Space and Upper Atmosphere Research Commission (Pakistan)
TIROS	Television Infra-red Observation Satellite (United States)
TM	Thematic Mapper (on board Landsat)
TOVS	TIROS Operational Vertical Sounder
TRSC	Thailand Remote Sensing Centre
UNDP	United Nations Development Programme
VSAT	Very small aperture terminal
WMo	World Meteorological Organization

Chapter 1

INTRODUCTION

1.1 Purpose, scope and structure of the inventory

The ESCAP Regional Space Applications Programme for Sustainable Development (RESAP) lays great emphasis on information dissemination and exchange. RESAP was formally launched at the Ministerial Conference on Space Applications for Development in Asia and the Pacific, held at Beijing in September 1994, through the adoption of the Beijing Declaration on Space Technology Applications for Environmentally Sound and Sustainable Development in Asia and the Pacific, the Strategy for Regional Cooperation in Space Applications for Sustainable Development and the Action Plan on Space Applications for Sustainable Development in Asia and the Pacific. The Beijing Declaration, Strategy and Action Plan were endorsed at the fifty-first session of ESCAP, held at Bangkok in May 1995, signifying the commitment of member and associate member countries to RESAP.

This inventory is part of the information dissemination and exchange effort of RESAP. ESCAP had published inventories on remote sensing in 1989 and 1990 under the aegis of its earlier Regional Remote Sensing Programme (RRSP). Much of that information is now outdated. This inventory on space technology capabilities, facilities and activities in the context of sustainable development is being published in response to the growing demand for information. Its publication is included in the RESAP work programme for 1997, which was endorsed by the Intergovernmental Consultative Committee (ICC) on RESAP at its second session held in June 1996.

The inventory provides information on pertinent aspects of space programmes in regional countries that are members of ICC on RESAP. It covers organizational matters, policies, facilities, investment, personnel, projects, training activities, seminars, workshops and conferences, databases and archives, and publications. Apart from the information forwarded by countries in response to a set of questionnaires, a synopsis of the national space effort in each country is given in the country profiles. Additionally, regional overviews in respect of national space programmes, as well as the four selected application areas, are included. These regional overviews and country profiles are based on a wide variety of sources, such as country reports and papers presented at various meetings, seminars and symposia organized by ESCAP or other agencies, newsletters, handbooks, journals, magazines and newspapers, and provide a useful general perspective of the overall situation. This inventory should be considered more like a concise survey on space technology capabilities in the region as used for sustainable development rather than a description of existing facilities/equipment alone.

The structure of the inventory corresponds to the framework devised for RESAP, under which attention is being focused on four selected "core" application areas, for which regional working groups have been established. The four application areas are (a) remote sensing, GIS and satellite-based positioning, (b) meteorological satellite applications and natural hazards monitoring, (c) satellite communication applications and (d) space sciences and technology applications.

1.2 Seeking information through questionnaires

Accordingly, four questionnaires corresponding to the four application areas were designed to seek information from respondent countries on a wide range of aspects, as mentioned above. To assist respondents in filling out the questionnaires, detailed instructions were prepared to obtain as much information of interest as possible in a standardized format. The questionnaires were discussed at the meeting of the national contact points (NCPs) for the RESAP Regional

Information Service (RIS) and Education and Training Network held in November 1995 and at the meetings of the four regional working groups held in the first half of 1996, and were improved in the light of these discussions. An additional questionnaire was prepared to solicit general information on overall national space programmes/activities, including organizational aspects, mandates, policies, training, coordination and promotional activities.

The set of five questionnaires was sent in October-November 1996 to the national focal points (NFPs), designated by their governments to represent their respective countries on ICC, as well as to the NCPs for RIS. Also, each of the four application-specific questionnaires were sent to the NCPs for the respective regional working groups. The countries targeted were those that are part of ICC on RESAP and the three-level network operating under RESAP. The NFPs on ICC constitute the first level of this network. The NCPs on the working groups form the second level, while the NCPs for RIS and education/training form the third level. Countries in this network are mainly those that are members of ICC on RESAP, which is, historically, the reconsituted version of the earlier ICC for RRSP. Membership of ICC is open to all members and associate members of ESCAP. Nevertheless, with a view to include information on space activities of non-ICC countries of the Asian and Pacific region, a short questionnaire seeking general information on space activities was distributed to representatives of these countries.

In general, regional countries that are part of the RESAP/ICC network have provided useful responses, on which the inventory is primarily based.

1.3 Outline and arrangement of chapters

Since it was intended that the inventory would consist mainly of inputs provided by countries in a standardized format, the filled-out questionnaires as received from the countries have been included, with minimal necessary editing, except in cases when it is most beneficial and more convenient to present the data at a regional level, such as investments in space technology facilities. It is to be noted that the absence of a country in a regional tabulation does not necessarily mean the absence in the country of whatever data is to be tabulated but rather that the contact point for that country did not include the information requested. In cases where countries have submitted detailed reports instead of completed questionnaires, the reports have been incorporated after being suitably abridged and edited; every attempt has been made to reflect, in the abridged versions, the essence of the original reports. In a few instances, countries have attached supplementary or complementary information. This additional information has been included, as deemed appropriate, after due editing.

The inventory is divided into seven chapters. Chapter 1 describes the purpose, scope and structure of the inventory, the preparation of questionnaires, the responses from countries and the arrangement of chapters, and adds some general comments on the usefulness of such publications.

Chapter 2 provides some background information enabling the reader to view the entire effort in the proper perspective. It provides an overview of space technology applications to sustainable development and summarizes the ESCAP space-related programmes in the regional context, including information dissemination and exchange activities.

Chapter 3 provides a regional overview of national space programmes and activities followed by the profiles of individual countries. The country profiles consist, in general, of two parts. The first is a synopsis of the overall national space effort based on information gleaned from country reports and papers presented at various meetings and seminars, as well as other sources, as mentioned under subsection 1.1 above. The second consists of the responses received to the questionnaire on national space programmes. Names, postal addresses, telephone and fax numbers, e-mail addresses and World Wide Web URLs (where available) of NFPs and responding persons are given so that further updated information may be obtained if desired. All countries that are part of the ICC/RESAP network are included since information about their space activities is available in ESCAP. While it was desired to include as many regional countries as possible, countries that are not ICC members had to be left out for want of sufficient information.

Chapters 4, 5, 6 and 7 are structurally similar to Chapter 3, and relate to (a) remote sensing, GIS and satellite-based positioning, (b) satellite meteorology and natural disaster monitoring, (c) satellite communication and distance education and (d) space sciences and technology applications, respectively. Each begins with a regional overview relating to the particular application area followed by individual country inputs focusing on programmes and activities in that application area, as provided by countries in the shape of the filled in question-naires. Names, postal addresses, telephone and fax numbers, e-mail addresses and Web URLs (if available) of NCPs and responding persons are given so that further updated information may be obtained if desired.

1.4 Some general comments

The relevance, validity and usefulness of most facts and figures, whether available in print or electronic media or both, is, in general, time-bound in the sense that with the passage of time the information becomes outdated. Timeliness of information is, therefore, of critical importance for operational use. Of course, information that has a value while it was still current retains a certain historical importance even after it has become outdated, especially for studying trends and patterns.

Despite the rapidly expanding role of the Internet and the increasing accessibility to information through cyberspace, the importance of the printed word as the physical counterpart and physical record of information is very much present. The role and significance of the printed media has not diminished – not yet at least. For one thing, many developing countries can provide this information (or any information, for that matter) only in print and have not yet put it on the Web. Also, the information on space technology and applications presently available on the Web would relate to the national space programmes of the respective individual countries. Information compiled in a particular format from different sources, giving a total picture and allowing direct comparisons between various aspects of the space programmes of different countries, as is being provided by this publication, would not be available from those Web pages. However, at some later stage, ESCAP may also be able to provide, through the Internet, such regional information as computerized regional databases, as recommended in the Strategy for Regional Cooperation and the Action Plan on Space Applications.

Chapter 2

BACKGROUND

2.1 Space technology for sustainable development

The year 1997 marked the fortieth anniversary of the launch, in 1957, of the first artificial satellite, Sputnik 1. It was a historic event which ushered in the Space Age and was the harbinger of momentous scientific and technological advances that were to follow in the coming years, but which could not have been fully perceived then. The space programmes have, on the one hand, added greatly to human knowledge about the Earth, the solar system and the universe, and, on the other, led to impressive advances in space and related technologies that have spawned an ever-proliferating number of applications and spin-off benefits having direct or indirect effects on human lives. Because of the stringent requirements imposed by the space environment, space programmes have acted as catalysts for the development of several science and engineering fields, such as, for instance, microelectronics, computer technology, materials science, cybernetics and biophysics, among others. Many of the advances made possible by space research have been truly astounding and have had far-reaching effects on a variety of human activities. The profound impact of space technology on communications, broadcasting, weather forecasting, resource and environmental management, navigation and positioning (to name a few) is well-known and part of common experience. A world bereft of the various satellite-based services and facilities routinely available today would be difficult to imagine and live in now. Indeed, space technology has transformed the world within a span of only one generation.

All scientific and technological advances can be viewed as part of the continuing economic, social and intellectual progress achieved by man over thousands of years since the dawn of civilization. This progress has been motivated by man's insatiable desire for a better life characterized by more wealth, comfort, power and security, and has been prompted by his irrepressible spirit of inquiry and adventure to understand and discover nature. Progress was initially slow, gathering increasing momentum with time. Until the advent of the Industrial Revolution about 200 years ago, human activities had had little impact, if any, on the environment and the natural cycles that sustain life on the Earth. However, the situation began to change with the Industrial Revolution as man's growing technological prowess and ability to modify his surroundings developed at an ever-accelerating pace. Over the last two centuries, he has been relentlessly exploiting natural resources, unconcerned about long-term sustainability, and changing, on a large scale, the natural environment, unaware of the possible degradation involved, in pursuit of economic or material development, entailing industrialization, urbanization, and expansion in commerce, communications, transportation and other infrastructure. In the process, mankind thoughtlessly indulged in environmentally damaging activities, such as emission of noxious gases and substances into the atmosphere from industries, fossil fuel burning in motor vehicles, discharge of poisonous industrial and municipal wastes into rivers and seas, over-irrigation, clearing of forests and natural wildlife habitats to make way for communication links or expanding cities, and more. The disastrous effects have manifested themselves in various ways, such as the greenhouse effect, stratospheric ozone depletion, acid rain, polluted seas and rivers, deforestation, soil salinization, desertification, destruction of ecosystems, loss of biodiversity, and a probable increase in floods, landslides, storms and other disasters. If timely remedial steps are not taken, the damage may be irreversible with serious consequences for mankind.

Natural processes transcend national boundaries, and environmental problems are often of global or regional dimensions. They can be effectively tackled only through concerted action at international or regional levels, with national environment conservation programmes of different countries working in synergy as components of an international or regional effort. Fortunately, environmental consciousness is growing everywhere and there is hope that through appropriate policies, more responsiveness among concerned sectors and the general public, effective actions using available scientific and technological means, and cooperation between relevant agencies within a country and between countries, environmental degradation may be arrested, and perhaps, eventually, reversed to some degree. It must be realized that development that disturbs the environmental or ecological balance is ultimately self-defeating and unsustainable. It is imperative that all countries adopt sustainable development, defined as "development that meets the needs of the present generation without compromising the ability of future generations to meet theirs", as the guiding principle in all socio-economic development activities, and cooperate with one another in optimizing resource and environment management. Development strategies should be so integrated with environment conservation strategies that sustainability is assured, as envisaged in Agenda 21 adopted by the United Nations Conference on Environment and Development at Rio de Janeiro in 1992.

Can space technology contribute to environmentally sound and sustainable development? The answer is most emphatically in the affirmative. Space technology can address many of the problems concerning environmental management identified in Agenda 21. In fact, space technology is ideally suited to support the integrated and holistic approach needed to tackle these problems of global or regional magnitude. In particular, satellite remote sensing data available, on a real-time or near-real-time basis, from a variety of visible, infra-red and microwave sensors at different spatial and spectral resolutions can provide the synoptic and repetitive information necessary for addressing resource and environmental problems in an integrated manner. Polar-orbiting and geostationary meteorological satellites provide data at high observational frequencies for routine weather forecasting, including disaster monitoring, and atmospheric research, as well as for non-meteorological uses like studying vegetation cover or the sea surface globally. Communication satellites provide data and voice communications, offer facilities for distance education, broadcasting and other uses, and relay information on impending natural disasters. Then there are specialized satellites or sensors providing data on various atmospheric parameters, ocean surfaces, the Earth's radiation budget, magnetic fields, and so on. Satellites also offer navigation and search-and-rescue services or provide precise position data for any point on the Earth. The applications of space technology are limited only by the human imagination. It is hardly surprising that international scientific research programmes, such as the International Geo-sphere-Biosphere Programme and World Climate Research Programme, designed to study the Earth as a single though complex integrated system, a view strengthened by space research itself, make extensive use of the increasing versatility of space technology and space-acquired data of the Earth.

2.2 The Asian and Pacific region and the space-related programmes of ESCAP

The ESCAP region covers most of the Asian continent, Australia and New Zealand, and most of the island countries and territories in the Pacific. This vast region, stretching from the Russian Federation in the north to New Zealand in the south and from French Polynesia in the east to Turkey in the west, comprises approximately a quarter of the world's land area and contains three-fifths its population. It is not surprising, therefore, that the region has tremendous diversity with regard to geography, race, language, religion, culture, political, economic and educational systems, with different countries at varying levels of economic, social, industrial and technological development. The region has some of the world's hottest and coldest places, wettest and driest areas, highest mountain ranges and flattest lowlands, most fertile agricultural lands and most barren deserts, most populous and least populous nations and most developed and least developed countries, to cite a few of its contrasts. The region consists mainly of developing countries, with a few exceptions. While some developing countries in East Asia have achieved impressive economic growth, most developing countries face common problems, such as rapidly expanding populations, poverty, illiteracy, poor health facilities, unplanned urban expansion, increasing pressure on resources and amenities, weak infrastructure and environmental deterioration in various forms such as deforestation, land degradation, pollution and other problems. The region is also vulnerable to disasters, such as floods, cyclones, landslides, earthquakes and tsunamis. There is a need to take effective measures to arrest the environmental damage and equip the region to cope better with natural calamities. Countries in the region must use space technology to facilitate environmental management, sustainable development, natural resource accounting and disaster monitoring, and should integrate space applications with economic development planning.

One regional programme designed to help countries in the region in this effort is RESAP, implemented by ESCAP since 1994. This programme is aimed at enhancing the capacity of developing countries in the region to use space technology for sustainable development. This is being achieved through various activities, such as conducting collaborative projects in selected application areas, organizing seminars, workshops and symposia, developing human resources through training programmes, including arranging fellowships, promoting information exchange and cooperation through networks and working groups, providing information, analyses, techniques, advisory services, and other activities. RESAP emphasizes an application-oriented rather than a "technology-push" approach for the region.

Earlier, ESCAP had been executing, since 1983, the Regional Remote Sensing Programme (RRSP), with UNDP funding support. Designed to develop the region's capabilities to use remote sensing for resource and environmental surveying and management, RRSP was the forerunner of RESAP. RRSP was considered a highly successful programme, which created a tangible impact on the remote sensing scenario in the region. Undoubtedly, its success provided the impetus to ESCAP to embark upon RESAP. Another project that immediately followed RRSP and was intended to consolidate and extend the achievements of RRSP was the project on integrated applications of geographic information systems and remote sensing for sustainable natural resources and environment management (GIS-RSRP). This project, which was initiated in 1993 with partial UNDP funding support, reflected the increasing emphasis on the use of GIS, along with remote sensing, in resource and environmental management. Both RRSP and GIS-RSRP have been absorbed as integral components of the regular activities of RESAP. The priority accorded to space applications by ESCAP is evident from the fact that the Space Technology Applications Section was established in the then Environment and Natural Resources Management Division of ESCAP in 1994. The Section serves as the nucleus for all space-related activities in ESCAP.

The Asian and Pacific region is better placed than other developing regions to absorb and use space technology for sustainable development. In general, regional countries have recognized space technology as an important enabling technology in the context of socio-economic progress. Several countries have made investments, commensurate with their resources, to enhance their capabilities in the domain of space and have developed fairly impressive facilities, including some that can launch satellites. As the United Nations' regional arm, ESCAP has supplemented these national efforts through the space-related regional collaborative programme it has been implementing since 1983.

2.3 Information exchange in the context of the space-related programmes of ESCAP

The dissemination and exchange of information and the sharing of experiences is an essential ingredient of RESAP. It is an underlying theme in the Strategy for Regional Cooperation in Space Applications and the Action Plan on Space Applications for Sustainable Development, on which RESAP is based, and runs like a common strand through all aspects of this programme.

The network formed under RESAP (and earlier under RRSP) plays a vital role in sharing information and experiences on space applications in the region. The country reports and other papers presented by NFPs, NCPs and others at ICC sessions, regional working group meetings and other seminars, workshops and meetings, organized under RESAP (or RRSP), as well as the formal and informal discussions at these meetings, are part of the continuing process of exchange of information, ideas and experiences. ESCAP has been publishing the proceedings of the various meetings it has organized. It has also published rosters of remote sensing projects and professionals, bibliographies, regional inventories of remote sensing facilities, reports of pilot projects and various studies as well as technical manuals and guidelines for users. In addition, ESCAP published, as background papers for the Ministerial Conference on Space Applications for Development in Asia and in the Pacific held at Beijing in September 1994, four comprehensive documents focusing on (a) status of space technology in the region, (b) issues concerning space technology applications for sustainable development in the region, (c) a strategy for regional cooperation in space applications and (d) an action plan on space applications, as well as a compendium on space technology and its applications to sustainable development, besides the proceedings of the Ministerial Conference and of a symposium organized in tandem with this event. Further, it has been bringing out, since 1983, the quarterly *Remote Sensing Newsletter,* re-named *Space Technology Applications Newsletter* in 1995, and, since 1988, the semi-annual *Asian-Pacific Remote Sensing Journal,* re-named *Asian-Pacific Remote Sensing and GIS Journal* in 1996. These publications collectively constitute a vast and invaluable reference source on space technology and remote sensing and their applications in the region, including the evolution and development, in a historical perspective, of space-related facilities and capabilities in the various countries over the last 14 years.

2.4 Earlier inventories under the Regional Remote Sensing Programme

As mentioned above, ESCAP had published regional remote sensing inventories under RRSP. Although an inventory is basically a list of facilities and equipment, the term was extended in this context to include information on remote sensing programmes, organizational aspects, personnel, projects, available data, training activities, coordination mechanisms and promotional activities, as well as facilities and equipment. These inventories have thus served the purpose of handbooks providing information on all aspects of interest concerning remote sensing in the region. The first such *Regional Remote Sensing Inventory* was published in May 1989. That publication covered remote sensing programmes in 16 countries in the region, containing information provided by countries in the prescribed format in response to a questionnaire. In December 1990, ESCAP published a supplement to the inventory, an equally comprehensive and voluminous document covering 18 countries, which was basically an updated version of the May 1989 inventory.

2.5 The present inventory

The present inventory is the first on space technology capabilities, facilities and activities to be published under the aegis of RESAP. It provides comprehensive information on pertinent aspects with emphasis on the use of space technology for sustainable development, which is the essential aim of RESAP. It can serve as the basis for initiating the creation of computerized regional databases on space technology applications, as recommended in the Strategy for Regional Cooperation and the Action Plan on Space Applications. The Strategy and Action Plan have also recommended the creation of such computerized databases at the national level.

Chapter 3

NATIONAL SPACE PROGRAMMES AND ACTIVITIES

3.1 Regional overview

3.1.1 Background

The Asian and Pacific region, barring a few developed countries such as, for instance, Australia and Japan, consists mainly of developing countries and areas at varying levels of development. It contains some least-developed countries, including those that are small island nations as well as those that are landlocked. While average economic growth rates in most developing countries in the region have been moderate, some developing countries in East Asia, such as, for instance, Malaysia, the Republic of Korea, Singapore and Thailand, have achieved remarkable economic growth over the last few years. For example, per capita incomes in Singapore now equal those in developed countries, while the Republic of Korea is steadily approaching that position. The region also has some countries, such as the Russian Federation, whose economies are in transition from being centrally planned to market economies.

Notwithstanding the impressive economic progress made by some countries, the region, as a whole, continues to experience widespread environmental degradation. This calls for effective remedial measures to be taken using all possible means, including space technology. It is necessary that the concept of sustainable development, which seeks to achieve an equilibrium between development needs and environmental considerations, is accepted not just in theory but also followed in practice in the countries of the region. There is a need to make still greater use of the power and versatility of space technology in support of sustainable development in the region. For this, it is necessary that space applications become an integral part of national economic development policy-making and planning.

3.1.2 Developments in national space programmes

In the last few years, there has been appreciable enhancement in the capabilities of the region in the domain of space. Most countries that were already pursuing space programmes have further upgraded their facilities, expanded and diversified into more sophisticated fields and new application areas, and further developed their human resources. Some of the other countries that were previously not involved in space-related activities have initiated space programmes. While four regional countries are space-faring, that is, having the capability to launch satellites, about 30 countries, including some of the least developed, landlocked and island nations, are now engaged in some sort of space programme. In many cases, the programmes are mainly focused on remote sensing applications. Collectively, over 20,000 scientists in the region are working on about 2,000 projects related to the use of remote sensing in resource and environmental management. There are over 1,000 GIS installations in the region, including many in government planning offices. About 11 ground receiving stations for acquisition and processing of satellite remote sensing data are operating in nine countries of the ESCAP region, which has more such stations than any other part of the world. These stations collectively receive data from about eight different remote sensing satellite systems. Also, there are a large number of stations that receive data from meteorological and environmental satellites for weather forecasting and disaster monitoring. Nine countries are involved in the development of small satellites for various applications. The region, as a whole, has access to about 20 international/ regional and national communication satellites, offering a growing market a variety of services. The region is well equipped to utilize the continuing advances in space science and technology.

The development of space capabilities in the various countries of the region has been uneven, in view of wide disparities in the size of the countries, their human, financial and technical resources, their priorities, aspirations and needs based on their overall national policies and political commitments, and their level of socio-economic, industrial and technological development. As a consequence, the region has, at one end, very advanced and comprehensive space programmes, encompassing sophisticated facilities and highly trained manpower, and at the other end, rudimentary programmes of very limited size and scope. Between the two extremes lies a wide spectrum of space programmes of varying size, scope and range.

3.1.3 Categorization of national space programmes in the region

The national space programmes of the 23 countries in the region that are part of the RESAP/ICC network (see table 3.1) can be classified into four categories, generally on the basis of the overall size, scope, diversity and sophistication of the programme. None of these categories is homogeneous, as there are significant variations in capabilities and achievements between countries in the same category, with different countries being strong in different fields or areas. The categorization should not be considered as being very rigid or "watertight", and some borderline cases could be placed in either of two adjacent categories. Also, there could be other ways of classifying the national space programmes in the region using different criteria.

The first category comprises those countries that are spacefaring and pursuing large and advanced space programmes covering both science and technology. There are four spacefaring countries in the region, namely China, India, Japan and the Russian Federation. Japan is one of the world's strongest economies and a leading country in the development and commercialization of technology, possessing highly sophisticated facilities and capabilities in all fields related to space. China and India, the world's two most populous countries, have been able to develop impressive capabilities in space technology, backed by strong space science support; they possess what is called an end-to-end indigenous space capability, right from the conceptualization of a satellite to its designing, fabrication, testing and launching, and they are pursuing comprehensive space applications programmes. Both countries are making extensive use of space technology to address many of the problems that they face. The Russian Federation, as the main inheritor of the space facilities and capabilities of the former Soviet Union, also possesses a wide array of advanced space capabilities as well as considerable experience and expertise.

The second category consists of countries with well-established and fairly wide-ranging and diversified programmes, often being executed by fully-fledged national space agencies. They do not yet have the end-to-end capability of the spacefaring nations, but have fairly advanced facilities and good expertise in several fields of space science, technology and applications, such as remote sensing and GIS applications, operation of ground receiving stations for satellite remote sensing data, satellite meteorology, disaster monitoring, neutral atmosphere research, ionospheric research, design of sensors, development of small satellites for Earth observations and communications (sometimes with technical collaboration of outside agencies), and applications of communication satellites, including distance education. Some examples of countries in this category are Australia, Indonesia, Pakistan and the Republic of Korea. Some countries in this category have their national communication satellite systems designed and launched under contract by leading companies in North America or Europe. Examples are Indonesia's Palapa, Malaysia's MEASAT, the Republic of Korea's Koreasat, and Thailand's Thaicom satellite systems. Several countries in this category are also engaged in, or about to initiate, programmes for development of small satellites for Earth observations, communications and other scientific purposes. Although Australia has been placed in this category, its case is perhaps unique in the sense that it had launched its own indigenously developed satellite from launching facilities situated on its territory as far back as 1967, but had later given up indigenous satellite launches. It now plans to develop such a capability again, so that it could provide launch services to others on a commercial basis. With its wide-ranging capabilities in both space

Table 3.1 National focal points of the Regional Space Applications Programme for Sustainable Development

Country	Name	Agency
Australia	Mr David L.B. Jupp	CSIRO Office of Space Science and Applications
Azerbaijan	Professor Arif Mekhtiev	Azerbaijan National Space Agency
Bangladesh	Mr M.A. Subhan	Space Research and Remote Sensing Organization
China	Ms Zheng Lizhong	National Remote Sensing Centre of China
Fiji	Mr M. Jaffar	Ministry of Lands, Mineral Resources and Energy
India	Mr K. Kasturirangan	Indian Space Research Organization
Indonesia	Professor Haryono Djojodihardjo	National Institute of Aeronautics and Space
Islamic Republic of Iran	Mr Ahad Tavakoli	Iranian Remote Sensing Centre, Ministry of Post, Telegraph and Telephone
Japan	Mr Eiichi Kawahara	Embassy of Japan (in Bangkok)
Malaysia	Mr Nik Nasruddin Mahmood	Malaysian Centre for Remote Sensing, Ministry of Science, Technology and the Environment
Mongolia	Mr Sodov Khudulmur	Mongolian National Remote Sensing Centre
Myanmar	Mr Htay Aung	Department of Meteorology and Hydrology
Nepal	Mr Rajendra B. Joshi	Forest Research and Survey Centre
New Zealand	Mr David Pairman	Landcare Research New Zealand, Ltd.
Pakistan	Mr Abdul Majid	Space and Upper Atmosphere Research Commission
Philippines	Ms Ida F. Dalmacio	Department of Science and Technology, Philippine Council for Advanced Science and Technology Research and Development
Republic of Korea	–	Ministry of Science
Russian Federation	Mr A.I. Medvedchikov	Russian Space Agency
Singapore	Mr Lim Hock	Centre for Remote Imaging, Sensing and Processing
Sri Lanka	Mr Sam Karunaratne	Arthur C. Clarke Centre for Modern Technologies
Thailand	Mr Jirapandh Arthachinta	National Research Council
Vanuatu	Mr Edwin Arthur	Ministry of Lands, Geology, Mines, Energy and Rural Water Supply
Viet Nam	Professor Tran Manh Tuan	National Centre for Science and Technology

science and space technology, Australia may well join the club of spacefaring nations in the foreseeable future.

The third category comprises countries whose space programmes are not very extensive, but which, nevertheless, possess reasonably good facilities and expertise in certain fields or areas on which the respective space programme is mainly focused. In most of these countries, the main emphasis is often on remote sensing and GIS applications and additionally on satellite meteorology, disaster monitoring or distance education in varying degrees. Most countries in this category usually have some kind of formal institutional arrangement, such as a space agency, remote sensing centre, or a designated department or organization functioning as the national focal point or coordinator for space activities. Examples of countries in this category are Bangladesh, Mongolia, Sri Lanka and Viet Nam.

The fourth category includes countries whose space programmes or activities are of limited scope and range. Examples of countries in this category from among the RESAP/ICC member countries are Nepal and Vanuatu. The reasons for the small size of the space programmes or the slow pace of development of the programmes in such countries are varied, and include financial and resource constraints, limited needs due to the small size of the country, little capacity to assimilate space technology because of inadequate infrastructure and human resources, or lack of awareness.

On the basis of whatever information is available from ESCAP, it appears that two non-ICC countries, namely, Kazakhstan, which has fairly extensive facilities and a broad-based space research programme, and Turkey, which has its own national satellite communications system called Turksat, could be placed in the second category. No non-ICC country can be included in the first category as none is spacefaring. Most non-ICC countries would thus be placed in the third and fourth categories. One could, perhaps, add a fifth category comprising countries that have virtually no space-related activity at all.

3.1.4 Organizational structure for national space programmes

Many countries of the region have national space agencies or national agencies responsible for coordination of space activities. In most such countries, these agencies normally function either under the country's chief executive, that is, the president or prime minister, through an appropriate controlling ministry, division or department or a high-powered council or committee, or, as is more often the case, under ministries concerned usually with science, technology, environment, natural resources or surveying. In a few cases, these agencies work under ministries dealing with defence, postal and telephone services and transport and communications. In several countries, the national remote sensing centres also function as national space agencies, space coordinating agencies or lead or focal agencies for space-related matters. The space programme in countries having national space or space coordinating agencies are formally structured with policy guidelines and directions being provided by the controlling authorities. The space programmes are centralized within the space agencies, which play a pivotal or lead role in the entire space endeavour. However, other agencies in the country are also involved, in varying degrees, in different facets of the programme. Also, some countries have established advisory or consultative committees, comprising representatives from various national user agencies, to provide advice and suggestions regarding the space programme from the perspective of a user, so as to make the space programme more responsive to user needs.

On the other hand, there are many countries in the region that do not have a national space agency or a national space coordinating agency as such. Space programmes and activities in such countries are obviously more decentralized, with different parallel organizations engaged in various aspects of space science, technology and applications and reporting to different controlling ministries or departments. Some overlap and duplication in such situations is unavoidable. Often, the leading or more active organizations in various fields or application areas become the nodal agencies for those fields in the country, in the absence of a single designated national space agency. However, in these countries, there is usually some central coordinating or steering mechanism, in the form of a coordination, advisory or consultative committee or body, to give overall direction to space activities.

National space programmes are, as a rule, implemented, funded and coordinated by national space and other government agencies. The field of satellite communications is generally an exception to the rule in the sense that it is run on commercial lines and is largely handled by the commercial sector. The reason for this is that satellite communications had attained commercial viability in the 1960s. The operation and management of international communication satellite systems, which provide the bulk of global communication services, are mainly in the hands of large intergovernmental organizations or consortia, such as Intelsat, that are run on commercial lines. The satellites are fabricated by aerospace and satellite manufacturers in the United

States of America or Europe, which are mostly private, and launched by organizations or entities that are government-owned or private or joint public-private sector enterprises. While some national satellite communications systems, such as China's Chinasat, are owned, operated and managed by government agencies, many national satellite communications systems, such as Malaysia's MEASAT or Thailand's Thaicom, are often owned and managed by companies that are private or joint public-private sector ventures, with the satellites being built by American and European companies. With the growing trend towards liberalization and privatization in the region, the telecommunications sector in many countries is opening up for private competitors. In several countries, there is a trend towards growing involvement of private service providers in operational activities, including satellite communications, with governments playing a regulating and supervisory role and being engaged more in planning activities. In almost all countries, satellite communications are invariably part of the overall telecommunications sector, the regulation of which is primarily the responsibility of ministries or departments concerned with communications. Thus, the field of satellite communications *per se* does not normally fall within the purview of activities handled by national space agencies, which generally function under ministries dealing with science, technology, the environment or surveying.

3.1.5 Developments relating to policy and institutional aspects of space programmes

National policies and institutional arrangements relating to the national space programme are critically important for the success of the programme. Policies define the objectives and parameters of the space programme, while institutional frameworks provide the mechanisms that enable effective implementation of the policies. It is a heartening sign that several countries in the Asia-Pacific region are taking the necessary steps in that direction. These steps include policy formulation, legislation, establishing new or suitably revised institutional arrangements and devising appropriate strategies. Some countries, such as Bangladesh, the Islamic Republic of Iran and Pakistan, have set up coordination or advisory committees at the national level, comprising representatives from relevant user and planning agencies as well as the space agency, to develop a synergy between the space and other sectors. This would lead to integration of space applications in development planning. In the Philippines a legislative bill has been passed to modernize and optimize resource and environment management through space technology. In Japan, new space policies have been formulated that stress an application-driven strategy and encourage international and regional cooperation in space technology development. Some countries, especially China and Mongolia, have formulated their respective national Agenda 21, in which space technology applications have been integrated with sustainable development planning. India has suitably expanded its national programmes relating to space applications for integrated environment and natural resource management, and it has offered to provide services for the region. Other countries, such as Fiji, Nepal and Viet Nam, have also begun to include space applications in their environment and development programmes.

3.1.6 Conclusion

The development of appropriate policies and institutional arrangements in countries will help in the integration of space applications in the national socio-economic development planning and implementation processes and lead to a significantly enhanced role for space technology in sustainable development and environmental management, as envisaged in the Strategy for Regional Cooperation and the Action Plan for Space Applications. Regional collaborative endeavours such as RESAP will act as catalysts in building national capacity in space technology for sustainable development and are, at the same time, expected to strengthen cooperation and goodwill among regional countries. It is believed that the region will be able to effectively tackle its resource and environment problems, and these are quite considerable, given its inherent dynamism and vitality and its improving capabilities in relevant fields, including space technology.

3.2   Country inputs

3.2.1 Australia

Australia has been involved in activities relating to space and atmospheric research since the 1960s. In fact, Australia was one of the earliest countries to launch an indigenously designed satellite from its own territory. This satellite, called WRESAT, was meant to collect scientific data of the upper atmosphere and was launched from a site in the country as far back as 1967. However, development of indigenous satellite launching capabilities was not pursued further at that stage.

A major restructuring and overhaul of the Australian national space programme took place in 1996. The Office of Space Science Applications (COSSA) of the Commonwealth Scientific and Industrial Research Organization (CSIRO) is responsible for the planning, execution and coordination of the space programme in the country in conjunction with other relevant departments, universities and industries, while the Space Policy Unit of the Department of Industry, Science and Tourism is to provide policy advice, keeping in view overall national interests. A Space Advisory Board, chaired by the Minister, meets periodically to assess the progress made and to provide guidelines for the future.

Australia's space programme is quite comprehensive and appropriately emphasizes both space science and space technology. This broad-based and diversified programme includes astronomy, astrophysics, upper and middle atmosphere research, solar physics, ionospheric physics, magnetospheric physics, remote sensing for resource mapping and environment monitoring, meteorology, climate modelling, global positioning, laser ranging, communications, sensor development and other allied fields. The work is carried out at a large number of universities spread across the country and government R and D organizations. As Australia is a developed country with strong R and D institutions, its contribution to research in space science and space technology and associated fields is quite significant. Some of the prominent R and D organizations in the country that are involved in space-related activities are CSIRO, the Australian Centre for Remote Sensing (ACRES) of the Australian Land Surveying and Land Information Group, the Bureau of Meteorology, the Bureau of Mineral Resources and the Antarctic Division.

COSSA also represents Australian interests in RESAP, as well as in the Committee on Earth Observation Satellites (CEOS) and the International Astronautical Federation (IAF), and coordinates with Australian industry in the context of space. The Australian Academy of Science is responsible for coordinating activities concerning the Committee on Space Research (COSPAR), including submission of the biennial national reports on space research to COSPAR. ACRES is managing and operating two ground receiving stations at Alice Springs, in the centre of the country, and at Hobart in Tasmania for acquisition of Landsat, SPOT, ERS, JERS and Radarsat data. Satellite remote sensing is of special relevance to Australia because of the need to survey and map this huge country and its natural resources, including abundant mineral resources, and monitor changes over such a vast area. Over 100 government organizations and a roughly equal number of research and academic institutions are making use of remote sensing and GIS techniques in the country.

Australia, which possesses good facilities and expertise in space science and technology and has an expanding industrial base, plans to develop an indigenous capability for satellite launching which could be offered to other countries on a commercial basis.

It is also planned to develop and launch a microsatellite for conducting sientific experiments in Earth observations and communications. To be called FedSat, the satellite is planned for launching in 2001 to coincide with the centenary of the establishment of the Federation.

The development and expansion of the telecommunications sector, including satellite communications, in the country is basically the responsibility of the private sector, as is the case now with an increasing number of countries. The Australian domestic satellite operator, Aussat, was privatized in 1991, under the name of Optus Communications. Under Aussat, a number of Aussat communication satellites were launched in the 1980s, some of which are still functioning. These, as well as other newer communication satellites, are now operated by Optus.

Further information on Australia's space programme may be obtained from the RESAP national focal point:

> Mr David L.B. Jupp
> Head and Science Program Leader
> CSIRO Office of Space Science and Applications
> GPO Box 3023
> Canberra ACT 2601
> Australia
> Fax: (616) 279-0812
> URL: http://www.cossa.csiro.au

General Information on National Space Activities

National space agency responsible for coordination of space activities	Commonwealth Scientific and Industrial Research Organization (CSIRO) Office of Space Science and Applications.
Mandate	To maximize the environmental, social and economic benefits to Australia arising from research and development in space-related science and engineering. For more information, refer to http://www.cossa.csiro.au/.
Organizational set-up	See figure 3.1.
Brief outline of the national space programme	Remote sensing, Earth observations, space science and astronomy, satellite communications, technology development.
Organizations responsible for various facets of the national space programme	Coordination: CSIRO Office of Space Science and Applications. Policy advice to government: Space Policy Unit, Department of Industry, Science and Tourism.
Legislation, government policy or directive related to integration of space applications in the development planning process	None.
Coordination and promotional activities	CSIRO Office of Space Science and Applications represents Australian interests in RESAP, in CEOS and in IAF. It coordinates with other Australian agencies in industry, with respect to this activity. The Australian Academy of Science coordinates scientific input relating to COSPAR. The Space Policy Unit, Department of Industry, coordinates matters related to COPUOS of the United Nations. In addition, there are several private interest groups promoting space activities in Australia, such as: • National Space Society of Australia • Australian Space Research Institute • Remote Sensing and Photogrammetry Association of Australia Major publications concerning space in Australia include: • CSIRO Space Industry News • ACRES Update • RS and PAA Newsletter • Australian Meteorological Magazine • Australia and New Zealand Physicist

**Status of space science
and technology education**

Several leading educational institutions are involved. These include:

- Space Science: La Trobe University, Newcastle University, University of Adelaide, Australian Defence Forces Academy. University of Tasmania, University of Sydney and others
- Remote Sensing and GIS: University of New South Wales, Curtin University
- Navigation and GPS: University of Technology (Queensland), Space Centre for Satellite Navigation
- Communications: University of Canberra, University of South Australia – Institute of Telecommunications Research

Regional space applications programme network

Contact	Name	Agency
RESAP national focal point	Mr David L.B. Jupp	CSIRO Office of Space Science and Applications
Regional Working Group on Remote Sensing, GIS and Satellite-based Positioning	Mr David L.B. Jupp	CSIRO Office of Space Science and Applications
Regional Working Group on Satellite Communication Applications	Professor Michael Miller	University of South Australia
Regional Working Group on Meteorological Satellite Applications and Natural Hazards Monitoring	Mr A.B. Neal	Bureau of Meteorology, Department of the Environment, Sport and Territories
Regional Working Group on Space Sciences and Technology Applications	Mr David Jauncey	CSIRO Office of Space Science and Applications

Respondent to questionnaires

Field	Agency	Name
Remote sensing, GIS and satellite-based positioning	CSIRO Office of Space Science and Applications	Mr Jeff Kingwell
Satellite communication applications	–	–
Meteorological satellite applications and natural hazards monitoring	Bureau of Meteorology	Mr A.B. Neal
Space sciences and technology applications	CSIRO Office of Space Science and Applications	Mr David Jauncey

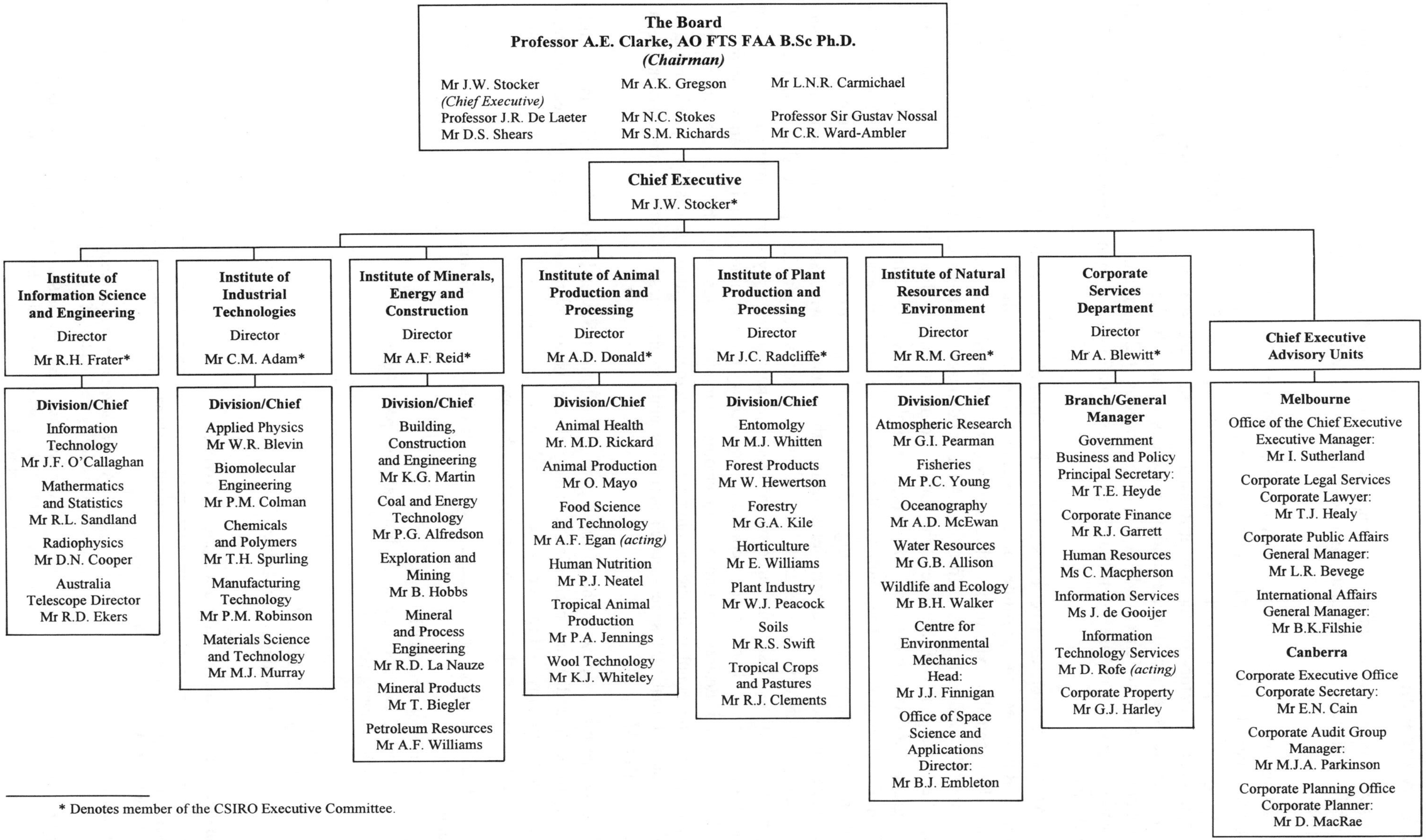

A

The Board
Professor A.E. Clarke, AO FTS FAA B.Sc Ph.D.
(Chairman)

Mr J.W. Stocker   Mr A.K. Gregson   Mr L.N.R. Carmichael
(Chief Executive)
Professor J.R. De Laeter   Mr N.C. Stokes   Professor Sir Gustav Nossal
Mr D.S. Shears   Mr S.M. Richards   Mr C.R. Ward-Ambler

Chief Executive
Mr J.W. Stocker*

Chief Executive Advisory Units

Melbourne
Office of the Chief Executive
Executive Manager:
Mr I. Sutherland
Corporate Legal Services
Corporate Lawyer:
Mr T.J. Healy
Corporate Public Affairs
General Manager:
Mr L.R. Bevege
International Affairs
General Manager:
Mr B.K. Filshie

Canberra
Corporate Executive Office
Corporate Secretary:
Mr E.N. Cain
Corporate Audit Group
Manager:
Mr M.J.A. Parkinson
Corporate Planning Office
Corporate Planner:
Mr D. MacRae

Corporate Services Department
Director
Mr A. Blewitt*

Branch/General Manager
Government Business and Policy
Principal Secretary:
Mr T.E. Heyde
Corporate Finance
Mr R.J. Garrett
Human Resources
Ms C. Macpherson
Information Services
Ms J. de Gooijer
Information Technology Services
Mr D. Rofe *(acting)*
Corporate Property
Mr G.J. Harley

Institute of Natural Resources and Environment
Director
Mr R.M. Green*

Division/Chief
Atmospheric Research
Mr G.I. Pearman
Fisheries
Mr P.C. Young
Oceanography
Mr A.D. McEwan
Water Resources
Mr G.B. Allison
Wildlife and Ecology
Mr B.H. Walker
Centre for Environmental Mechanics
Head:
Mr J.J. Finnigan
Office of Space Science and Applications
Director:
Mr B.J. Embleton

Institute of Plant Production and Processing
Director
Mr J.C. Radcliffe*

Division/Chief
Entomology
Mr M.J. Whitten
Forest Products
Mr W. Hewertson
Forestry
Mr G.A. Kile
Horticulture
Mr E. Williams
Plant Industry
Mr W.J. Peacock
Soils
Mr R.S. Swift
Tropical Crops and Pastures
Mr R.J. Clements

Institute of Animal Production and Processing
Director
Mr A.D. Donald*

Division/Chief
Animal Health
Mr. M.D. Rickard
Animal Production
Mr O. Mayo
Food Science and Technology
Mr A.F. Egan *(acting)*
Human Nutrition
Mr P.J. Neatel
Tropical Animal Production
Mr P.A. Jennings
Wool Technology
Mr K.J. Whiteley

Institute of Minerals, Energy and Construction
Director
Mr A.F. Reid*

Division/Chief
Building, Construction and Engineering
Mr K.G. Martin
Coal and Energy Technology
Mr P.G. Alfredson
Exploration and Mining
Mr B. Hobbs
Mineral and Process Engineering
Mr R.D. La Nauze
Mineral Products
Mr T. Biegler
Petroleum Resources
Mr A.F. Williams

Institute of Industrial Technologies
Director
Mr C.M. Adam*

Division/Chief
Applied Physics
Mr W.R. Blevin
Biomolecular Engineering
Mr P.M. Colman
Chemicals and Polymers
Mr T.H. Spurling
Manufacturing Technology
Mr P.M. Robinson
Materials Science and Technology
Mr M.J. Murray

Institute of Information Science and Engineering
Director
Mr R.H. Frater*

Division/Chief
Information Technology
Mr J.F. O'Callaghan
Mathematics and Statistics
Mr R.L. Sandland
Radiophysics
Mr D.N. Cooper
Australia Telescope Director
Mr R.D. Ekers

---

* Denotes member of the CSIRO Executive Committee.

Figure 3.1 Organizational structure of CSIRO (as at 1 July 1993).

3.2.2 Azerbaijan

Azerbaijan was one of the 15 constituent republics of the former Union of Soviet Socialist Republics. It became a separate and sovereign country following the dissolution of the USSR in 1991.

Its national space agency, the Azerbaijan National Aerospace Agency (ANASA), has been involved, since 1975, in various aspects of the erstwhile Soviet space programme. For example, an X-ray telescope developed by ANASA during the Soviet era still functions in the astrophysical laboratory on the Soviet/Russian manned space station, Mir. ANASA was involved in various fields at that time, including remote sensing. It thus possesses facilities, as well as personnel, with experience and expertise in various fields relating to space technology and its applications, which are contributing in different nation-building activities in an independent Azerbaijan.

ANASA comprises five constituent units, namely the Institute for Space Research of Natural Resources, the Institute of Aerospace Informatics, the Ecology Institute, the Special Design Unit and the Experimental Industry Work. Each of these units is responsible for specific aspects of the national space effort. ANASA is also actively engaged in collaborative activities with the Academy of Sciences, the State Committees on Hydrometeorology, Geodesy and Cartography and on Ecology of Azerbaijan Republic as well as other government, research and industrial organizations. The main thrust areas in the national space programme are the use of satellite and airborne remote sensing and GIS technologies to study natural resources and the environmental and ecological problems facing the country, with special focus on the Caspian Sea area. ANASA is also engaged in developing software for data processing and thematic mapping as well as designing airborne information measurement systems comprising sensor and control subsystems. Further, it is contemplating setting up a ground receiving station for remote sensing data and desires to improve electronic communications and the state of information technology in the country in the coming years. Azerbaijan is also increasingly participating in regional and international space-related activities and programmes. It acts as the national focal point for RESAP.

Like the other republics that formed part of the former Soviet Union, Azerbaijan is currently passing through a difficult period as it makes the transition from a centrally planned economy to a market economy. However, a turnaround in the economic situation should help to enlarge the scope and size of the national space effort beyond its current modest level.

Further information on Azerbaijan's space programme may be obtained from the RESAP national focal point:

> Professor Arif Mekhtiev
> General Director, Azerbaijan National Aerospace Agency
> 159, Azadlyg pr.
> Baku 370106
> Azerbaijan
> Fax: (89 22) 621-738
> E-mail: mekhtiev@anasa.baku.az

General Information on National Space Activities

National space agency responsible for coordination of space activities	Azerbaijan National Aerospace Agency (ANASA).
Mandate	ANASA is a government organization with a mandate specified in a decree issued by the President of Azerbaijan.
Organizational set-up	ANASA consists of Institute for Space Researches of Natural Resources, Institute of Aerospace Informatics, Ecology Institute, Special Design Unit and Pilot Works.

ANASA is also actively involved in joint and cooperative activities with the Academy of Sciences, State Committees on Hydrometeorology, Geodesy and Cartography, and Ecology of Azerbaijan and other government, research and industrial organizations.

Brief outline of the national space programme	Development and creation of nationwide system for ecology monitoring, including GIS, natural and anthropogenic environment pollution, natural disasters, land cover and its degradation, climate and weather. Development and creation of system for collecting, processing, archiving and dissemination of aerospace data.
Organizations responsible for various facets of the national space programme	Academy of Sciences, State Committee on Hydrometeorology, State Committee on Geodesy and Cartography and State Committee on Ecology.
Legislation, government policy or directive related to integration of space applications in the development planning process	Decree by Council of Ministers of Azerbaijan Republic No. 370 dated 7 September 1989.
Coordination and promotional activities	ANASA is responsible for coordinating activities carried out in connection with the national space programme by other relevant organizations, including the Academy of Sciences, Ministry of Communications, Ministry of Agriculture, Ministry of Economy, State Committees on Hydrometeorology, Geodesy and Cartography, and Ecology.
Status of space science and technology education	ANASA has a specialized department in Azerbaijan Oil Academy (University), joint research/teaching laboratory with Technical University.
	ANASA itself delivers courses on M.Sc. and Ph.D. degree programmes and is also actively involved in the Centre of People Teaching and Training for Behaviour during natural catastrophes.

Regional space applications programme network

Contact	Name	Agency
RESAP national focal point	Professor Arif Mekhtiev	Azerbaijan National Aerospace Agency (ANASA)
Regional Working Group on Remote Sensing, GIS and Satellite-based Positioning	–	–
Regional Working Group on Satellite Communication Applications	–	–
Regional Working Group on Meteorological Satellite Applications and Natural Hazards Monitoring	–	–
Regional Working Group on Space Sciences and Technology Applications	–	–

Respondent to questionnaires

Field	Agency	Name
Remote sensing, GIS and satellite-based positioning	Azerbaijan National Aerospace Agency (ANASA)	Mr Rustam B. Rustamov
Satellite communication applications	Azerbaijan National Aerospace Agency (ANASA)	Mr Rustam B. Rustamov
Meteorological satellite applications and natural hazards monitoring	Azerbaijan National Aerospace Agency (ANASA)	Mr Rustam B. Rustamov
Space sciences and technology applications	Azerbaijan National Aerospace Agency (ANASA)	Mr Rustam B. Rustamov

3.2.3 Bangladesh

The space programme in Bangladesh is executed by the Bangladesh Space Research and Remote Sensing Organization (SPARRSO), established in 1980. Broadly in line with the recommendations of the Ministerial Conference on Space Applications for Development, held at Beijing in September 1994, Bangladesh has set up a national policy-making committee involving various ministries, departments and universities to advise on the national space applications programme.

Remote sensing is the mainstay of the national space programme, in view of its great relevance to the needs of the country. As one of the less developed countries of the region, characterized by overpopulation, poverty, growing pressure on land, natural resources and civic amenities, inadequate infrastructure and environmental degradation, with a high frequency of occurrence of natural disasters, Bangladesh became involved in remote sensing soon after the launch of the first dedicated Earth resources satellite, Landsat 1, in 1972. Through cooperative projects with UNDP and NASA, sophisticated facilities for reception of data from meteorological satellites like NOAA and GMS were established along with facilities for Landsat data analysis. With the passage of time, the remote sensing data analysis facilities were upgraded and GIS facilities were added to complement remote sensing data analysis.

With the availability of synoptic and repetitive satellite images, it has become possible for concerned agencies in the country to closely monitor the build-up and movement of cyclones, tornadoes, storms and other weather systems and issue warnings well in advance of the disasters. This enables people to be evacuated to safer places in good time, thereby helping to save lives. Before satellite images began to be regularly used in Bangladesh for closely monitoring these disastrous weather systems, the toll of human lives resulting from such calamities was always much heavier. This reduction in the loss of human life as a result of cyclones and storms is a direct and tangible benefit of the application of space technology.

Satellite remote sensing data are being used in Bangladesh for a wide range of applications such as agricultural crop assessment, mapping of mangrove forests, monitoring mangrove afforestation, flood plain mapping, monitoring of floods, land-cover and land-use classification, and environmental monitoring. A wide cross-section of user departments and organizations in the country, besides SPARRSO, are making use of remote sensing data and GIS in various projects. These organizations include the Bangladesh Meteorology Department, the Water Resources Board, the Forest Department and the Bangladesh Bureau of Statistics. In view of extensive cloud cover over the country during the summer monsoon season, SPARRSO is initiating studies employing the cloud-penetrating SAR data from Radarsat to study the jute crop and the floods.

The Bangladesh Telegraph and Telephone Board operates all the satellite communication systems in the country. A distance education programme has been implemented as a regular programme of the Bangladesh Open University. Information superhighway links have also been established to make use of modern communication media.

Further information on the space programme of Bangladesh may be obtained from the RESAP national focal point:

Mr M.A. Subhan
Chairman, SPARRSO
Mohakash Biggyan Bhavan
Agargaon, Sher-E-Bangla Nagar
GPO Box 529, Dhaka 1207
Bangladesh
Fax: (88 02) 813-080
E-mail: sparrso@bangla.net

General Information on National Space Activities

National space agency responsible for coordination of space activities	Bangladesh Space Research and Remote Sensing Organization (SPARRSO).
Mandate	To apply space technology in surveying natural resources and monitoring the environment and natural disasters in the country; to establish data acquisition, processing and dissemination system (satellite ground station), develop instrumentation facilities and trained manpower; to act as the national focal point for space and remote sensing activities in the country and to provide the government with relevant information in formulating national, regional and international policy issues concerning applications towards sustainable development and to establish regional and international cooperation and collaboration in the peaceful uses of space science and technology.
Organizational set-up	As per organization chart in figure 3.2, SPARRSO has linkages with various relevant government and semi-government agencies dealing with remote sensing data applications. It also has linkages with various universities and academic institutions and supports R and D activities.
Brief outline of the national space programme	Installing and maintaining ground segments to receive, archive and process remote sensing data; carrying out R and D work in various sectors like agriculture, forestry, fisheries, water resources, meteorology, environment, soil science, cartography, geology, oceanography, and others; and monitoring of disasters in the country.
Organizations responsible for various facets of the national space programme	SPARRSO, the national space agency, is mainly responsible for the implementation of the national space programme.
Legislation, government policy or directive related to integration of space applications in the development planning process	The Government of Bangladesh has constituted a national committee, in line with the recommendations of the Beijing Declaration to oversee policy issues concerning space applications.
Coordination and promotional activities	SPARRSO disseminates information about its activities to user agencies by publishing annual reports and quarterly newsletters and by arranging training courses, seminars, visits, etc. It also provides advisory services to the user agencies and takes up joint research projects with them for wider application of the technology in various fields.
Status of space science and technology education	Space science and technology education has been included in the curriculum of some of the universities in the country. Students also carry out research in SPARRSO under joint supervision. SPARRSO scientists also obtain higher degrees from local universities.

Regional space applications programme network

Contact	Name	Agency
RESAP national focal point	Mr M.A. Subhan	Bangladesh Space Research and Remote Sensing Organization
Regional Working Group on Remote Sensing, GIS and Satellite-based Positioning	Mr Anwar Ali	Bangladesh Space Research and Remote Sensing Organization
Regional Working Group on Satellite Communication Applications	–	–
Regional Working Group on Meteorological Satellite Applications and Natural Hazards Monitoring	Mr A.M. Choudhury	Bangladesh Space Research and Remote Sensing Organization
Regional Working Group on Space Sciences and Technology Applications	Mr D.A. Quader	Bangladesh Space Research and Remote Sensing Organization

Respondent to questionnaires

Field	Agency	Name
Remote sensing, GIS and satellite-based positioning	–	–
Satellite communication applications	–	–
Meteorological satellite applications and natural hazards monitoring	Bangladesh Space Research and Remote Sensing Organization	Mr A.M. Choudhury
Space sciences and technology applications	Bangladesh Space Research and Remote Sensing Organization	Mr Nazmul Hoque

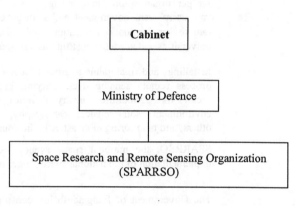

Figure 3.2 Organization of space activities in Bangladesh

3.2.4 China

China shares with India, Japan and the Russian Federation the distinction of being one of the four spacefaring countries in the ESCAP region. As the world's most populous country, with large and varied natural resources, a rapidly growing economy, expanding industry, improving infrastructure and a huge reservoir of qualified and trained manpower, China is pursuing a large, ambitious and wide-ranging national space programme. The country possesses adequate indigenous capabilities and facilities to execute this large programme covering both the space and the ground segments.

Owing to the wide-ranging scope and scale of the national space programme and the large size of the country itself, a large number of government departments and organizations, with complex interrelationships and linkages, are involved in different aspects and facets of the space programme (see figure 3.3). Under the apex State Council, there are two principal bodies, the State Science and Technology Commission of China (SSTCC) and the Commission of Science, Technology and Industry for National Defence. Under each of these two bodies, various departments and organizations function, dealing with specific fields. Some of the important organizations concerned with space-related activities are the Chinese Academy of Sciences (CAS), the China National Space Administration, the National Remote Sensing Centre and the Ministry of Posts and Telecommunications. Associated with or under each of these organizations there are, in turn, a large number of other institutions, centres and units. Many universities in the country are also engaged in research activities concerned with space. The very large number of organizations and institutions involved in space-related activities can be judged from the fact that more than 450 different agencies in the country are engaged in one way or another in remote sensing technology and its applications.

China is involved in a wide spectrum of research activities concerning space science and applications. These include, but are not limited to, astronomy, astrophysics, interplanetary research, solar-terrestrial physics, magnetospheric and ionospheric physics, remote sensing for resource and environmental surveying and management, middle and upper neutral atmosphere research, microgravity research and its applications, and life and biological sciences. This wide-ranging research work is carried out at premier research organizations like CAS or the Centre for Space Science and Applied Research (CSSAR) as well as the universities.

The country has also made significant strides in space technology. It has developed its own launchers, the most advanced and capable being the Long March expendable vehicles which have been used over the last 12 years or so to launch its own satellites as well as those that belong to other countries. China launched its first experimental satellite in 1967. Since then, it has designed and launched over 30 satellites, about half of which have been recoverable. These satellites include both experimental and operational ones for a wide range of applications in meteorology, disaster monitoring, remote sensing for resource and environmental management, communications, distance education, microgravity research and scientific data collection and experimentation for various purposes. Among these are the various scientific satellites dedicated to specific scientific purposes, such as DQ-1A and DQ-1B for studying upper atmosphere density or SJ-4 for studying the distribution and energy spectrum of charged particles in space around the Earth and their effects on spacecraft and satellite performance. China also has its own operational polar-orbiting meteorological satellites of the FY-1 series. It has recently launched the first of its FY-2 series of operational geostationary meteorological satellites. China has had its own operational communication satellites since 1988, apart from being linked to the global Intelsat system. Further, an operational Earth resources/remote sensing satellite is currently under development as a collaborative venture between China and Brazil.

The country has been operating its ground receiving station at Beijing for acquisition and processing of satellite remote sensing data since 1986. The station is presently capable of

23

acquiring and processing data from Landsat-5, ERS-1/2 and JERS-1, and is expected to begin Radarsat and SPOT data reception by the end of 1997. In addition, there are several ground stations for receiving data from various meteorological satellite, such as China's own FY-1 and FY-2 satellites, the American NOAA and Japanese GMS satellites, as well as Earth stations for communication satellites.

China is contributing to international and regional space programmes both bilaterally and multilaterally. It has substantial interaction with ESCAP, having organized the first Ministerial Conference on Space Applications for Development in 1994 to launch RESAP. It offers fellowships annually to trainees selected by ESCAP from developing countries to study remote sensing and GIS, surveying and mapping at the country's leading university in this field.

Further information on China's space programme may be obtained from the RESAP national focal point:

Ms Zheng Lizhong
Deputy Director-General
National Remote Sensing Centre of China
15B Fuxing Road
Beijing 100862
China
Fax: (86 10) 6851-3212

Regional space applications programme network

Contact	Name	Agency
RESAP national focal point	Ms Zheng Lizhong	National Remote Sensing Centre of China, SSTCC
Regional Working Group on Remote Sensing, GIS and Satellite-based Positioning	Ms Zheng Lizhong	National Remote Sensing Centre of China, SSTCC
Regional Working Group on Satellite Communication Applications	Mr Wang Yan Guang	Beijing Institute of Satellite Information Engineering, CAST
Regional Working Group on Meteorological Satellite Applications and Natural Hazards Monitoring	Professor Xu Jianmin	National Satellite Meteorological Centre of China
Regional Working Group on Space Sciences and Technology Applications	Mr Sun Huixian	Centre for Space Science and Technology Applications, CAS

Respondent to questionnaires

Field	Agency	Name
Remote sensing, GIS and satellite-based positioning	National Remote Sensing Centre of China, Changsha Branch	Mr Ni Jian-Ping
	Remote Sensing of Forestry, Chinese Academy of Forestry Sciences	Mr Zhao Xian Wen
	Wuhan Technical University of Surveying and Mapping	Mr Li Qingquan
Satellite communication applications	–	–
Meteorological satellite applications and natural hazards monitoring	National Satellite Meteorological Centre of China	Professor Xu Jianmin
	Institute of Remote Sensing Applications	Mr Zhang Jianzhong
Space sciences and technology applications	–	–

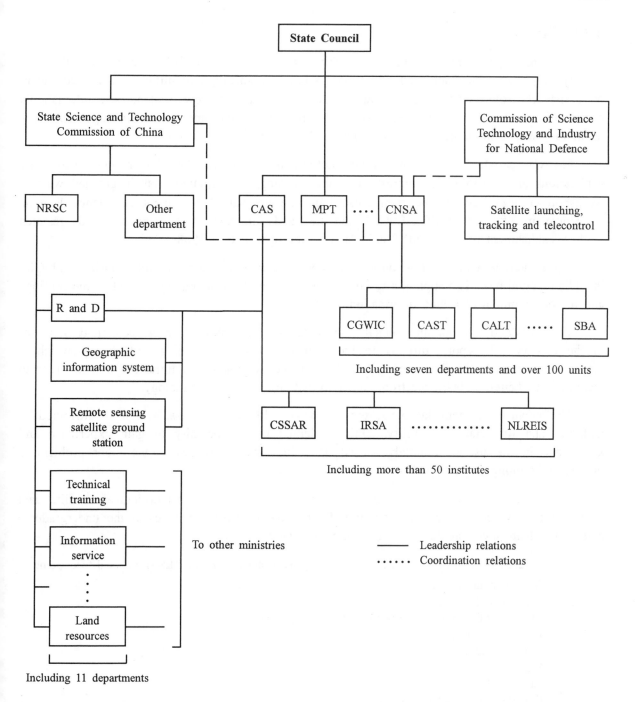

C

NRSC — National Remote Sensing Centre
CAS — Chinese Academy of Sciences
MPT — Ministry of Post and Telecommunications
CNSA — China National Space Administration
CSSAR — The Centre for Space Science and Technology Research
IRSA — Institute of Remote Sensing Applications
NLREIS — National Labour for Resources and Environment Information System
CGWIC — China Great-Wall Industry Corporation
CAST — Chinese Academy of Space Technology
CALT — Chinese Academy of Launch Vehicle Technology
SBA — Shanghai Bureau of Astronautics

Figure 3.3 Organization of space activities in China

3.2.5 Fiji

The space programme in Fiji, an archipelago of more than 300 islands, is still in its infancy. Fiji, like other Pacific island nations, is not economically self-sufficient and is dependent on assistance from neighbouring countries like Australia and New Zealand and international organizations such as the United Nations. Thus, the government tends to approach new technology with caution and demands strong justification for its application.

Fiji was satisfied with using aerial photographs as source data for mapping and other applications until 1986 when interest in remote sensing, in particular, was heightened when a workshop co-organized by the then Australian International Development Assistance Bureau, the University of South Pacific, ESCAP and the Lands and Survey Department directly resulted in the establishment of the RRSP South Pacific subprogramme in 1988.

Since then, a number of activities have been carried out using remote sensing and GIS, including natural hazards mapping, a resource inventory, coastal zone mapping, satellite meteorology and the development of a land information system.

The Fiji Land Information Council coordinates and promotes all space technology activities in the country. It reports directly to the Ministry of Lands, Mining and Energy which is mandated to provide the data, technology, human and administration infrastructures in order to maximize the benefits obtainable from land information and space technology.

The greatest problem faced, at present, is satellite data access, which is attributed to isolation and financial incapability. Fiji is denied the opportunity to gain experience and knowledge in the use of remotely sensed data for varied purposes. One other problem is the lack of training of personnel.

These problems will be addressed with the installation of a meteorological satellite data reception facility and the introduction of GIS and remote sensing courses at the undergraduate and postgraduate levels at the University of South Pacific.

Further information on Fiji's space programme may be obtained from the RESAP national focal point:

Mr Mohammed Jaffar
Permanent Secretary
Ministry of Lands, Mineral Resources and Energy
Government Buildings
P.O. Box 2222, Suva
Fiji
Fax: (67 9) 304-037
E-mail: frupendi@lands.govt.fi

General Information on National Space Activities

National space agency responsible for coordination of space activities	Ministry of Lands, Mining and Energy.
Mandate	To provide the data, technology, human and administration infrastructures to maximize the benefits obtainable from land information and space technology.
Organizational set-up	The focal point for space activities is the Ministry of Lands, Mining and Energy. Other agencies involved are the Fiji Land Information Council, the National Coordinating and Management Forum, the Fiji Land Information Support Centre and two liaison groups, namely the GIS and Remote Sensing Interest Group and Urban Utilities Group.
Brief outline of the national space programme	National mapping activities are carried out by the Department of Lands and Surveys. A national forest inventory is being carried out by

the Department of Forestry. Relevant education and training activities are carried out by the University of South Pacific (USP). Land use and capability mapping activites are carried out by the Agriculture Department. National meteorological services are profided by the Department of Meteorology.

Organizations responsible for various facets of the national space programme	Ministry of Agriculture, Fisheries and Forests; University of the South Pacific; Department of Meterology, etc.
Legislation, government policy or directive related to integration of space applications in the development planning process	None.
Coordination and promotional activities	The Fiji Land Information Council is the coordinating and policy-making body. The Council reports to the Minister for Lands, Mining and Energy. Below the Council is the Fiji Land Information Support Centre, the action group. Two formal liaison forums have been established, the GIS and Remote Sensing Interest Group and the Urban Utilities Group. The Council and both liaison groups meet once a month.
Status of space science and technology education	Human resources development is critical considering that GIS, LIS and remote sensing applications and implementation are still in their infancy in Fiji. For the first time, USP will commence a Certificate and Diploma courage in GIS and remote sensing from January 1997. As users become more familiar with these technologies, expectations will become high; therefore, there is a growing demand for greater use of GIS and remote sensing.

Regional space applications programme network

Contact	Name	Agency
RESAP national focal point	Mr M. Jaffar	Ministry of Lands, Mining and Energy
Regional Working Group on Remote Sensing, GIS and Satellite-based Positioning	Mr M. Jaffar	Ministry of Lands, Mining and Energy
Regional Working Group on Satellite Communication Applications	Mr Bruce Davis	University of the South Pacific
Regional Working Group on Meteorological Satellite Applications and Natural Hazards Monitoring	Mr Rajendra Prasad	Ministry of Tourism and Civil Aviation
Regional Working Group on Space Sciences and Technology Applications	Mr Les Allinson Mr Rupenl Anise	SOPAC Ministry of Agriculture, Fisheries, Forests and ALTA

Respondent to questionnaires

Field	Agency	Name
Remote sensing, GIS and satellite-based positioning	Ministry of Lands, Mining and Energy University of South Pacific	Mr Kemueli Masikirei Mr Bruce Davis
Satellite communication applications	–	–
Meteorological satellite applications and natural hazards monitoring	–	–
Space sciences and technology applications	–	–

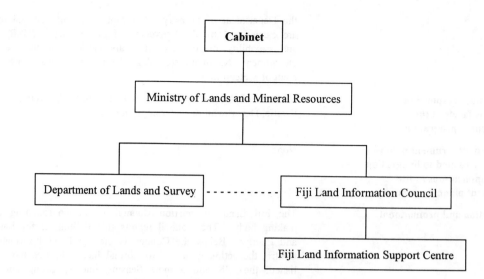

Figure 3.4 Organization of space activities in Fiji

3.2.6 India

India is one of the four spacefaring countries in the ESCAP region. It is the world's second most populous country with well-established educational institutions and science and technology R and D organizations, a large number of qualified scientists and technologists, and a steadily growing and increasingly diversified industrial base. India is engaged in an expanding and ambitious space programme and possesses the facilities and capabilities required to support the space and ground segments of this programme.

The Government of India established a Space Commission and separate Department of Space directly under the Prime Minister in 1972 exclusively for space matters (see figure 3.5). The Space Commission formulates overall policies regarding space, while the Department of Space is responsible for the actual implementation of the national space programme through the Indian Space Research Organization (ISRO). Traditionally, it has been the practice that the Secretary (that is, the functional head) of the Department of Space is also the Chairman of ISRO. Under the Department of Space there are, besides ISRO, other organizations such as the National Remote Sensing Agency (NRSA), the Physical Research Laboratory and the National Mesosphere-Stratosphere-Troposphere Facility, all of which work closely with ISRO. A comprehensive programme called the National Natural Resources Management System using satellite remote sensing data in selected priority areas has been under way for several years. NRSA has set up five major regional remote sensing centres besides the main centre. Under ISRO, several centres/units dedicated to various specific fields are functioning. Also, a large number of other government R and D institutions, various semi-government organizations, private sector agencies and universities across India are engaged in space-related research and applications work. This is particularly true in the case of satellite remote sensing where a large number of agencies, including remote sensing centres established by the various state governments, are involved in the application of the data.

The development of indigenous expertise and capabilities has been a cornerstone of India's space policy, and much of the space effort has relied increasingly on the country's own technical and human resources. When viewed against this perspective in particular, India's achievements in the domain of space are quite impressive. The Indian space programme is one of the largest in Asia and in the developing world as a whole. It is broad-based and comprehensive, covering diverse areas of space science, technology and applications, which include, but are not limited to, astronomy, astrophysics, cosmic rays, middle and upper atmosphere research, ionosphere and magnetosphere research, radio wave propagation, solar physics, remote sensing and its applications to resource and environmental surveying and management, satellite meteorology, design and development of different types of operational remote sensing, meteorological and communication and experimental and scientific satellites, subsystems, sensors and components, sounding rockets and launch vehicles for launching satellites, as well as establishment of ground receiving stations.

After the launch of its first experimental satellite, Aryabhata, and other early experimental satellites like Bhaskara I and II in the 1970s and early 1980s, India went on to design and develop bigger and more versatile satellites, some of which are providing operational service in such vitally important areas as Earth observation for resource mapping and environmental monitoring, weather forecasting and disaster monitoring and mitigation, communications and distance education. The earlier communication-cum-meteorological satellites of the INSAT series (INSAT-1) were built by leading American companies on contract, and also launched by foreign launchers. The currently operational INSAT-2A and B were developed indigenously but launched by foreign rockets. The IRS series of remote sensing satellites have also been developed indigenously. The latest, IRS-1C, provides a spatial resolution of about 6 metres in the panchromatic mode, which is better than that of any other current operational civilian satellite system. EOSAT of the United States markets IRS data worldwide. The IRS-P2 and

P3 were launched using India's own launcher, the Polar Satellite Launch Vehicle, putting India among the countries having satellite launch capabilities. The larger Geostationary Satellite Launch Vehicle is now to be tested over the next few years. India has been operating a ground station for reception of remote sensing data at Hyderabad since 1979. There are also a number of ground stations for receiving data from meteorological satellites as well as Earth stations linked to international, regional and national satellite communications systems.

A Centre for Space Science and Technology Education for Asia-Pacific was established at Dehra Dun in India in 1995, as part of United Nations sponsored efforts to set up such centres in different regions of the world. More nodes under such arrangements may be set up later in other countries of the vast Asian and Pacific region. The Centre presently conducts courses in remote sensing, satellite meteorology and satellite communications for professionals from the region. It also offers fellowships to trainees selected by ESCAP.

Further information on India's space programme may be obtained from the RESAP national focal point:

Mr K. Kasturirangan
Secretary, Department of Space, and Chairman, Indian Space Research Organization
Antariksh Bhavan
New Bel Road
Bangalore 560 094
India
Fax: (91 80) 341-5298, 341-2471
E-mail: krangan@isro.ernet.in

Regional space applications programme network

Contact	Name	Agency
RESAP national focal point	Mr K. Kasturirangan	Indian Space Research Organization, Department of Space
Regional Working Group on Remote Sensing, GIS and Satellite-based Positioning	Mr V. Jayaraman	Indian Space Research Organization, Department of Space
Regional Working Group on Satellite Communication Applications	–	–
Regional Working Group on Meteorological Satellite Applications and Natural Hazards Monitoring	–	–
Regional Working Group on Space Sciences and Technology Applications	–	–

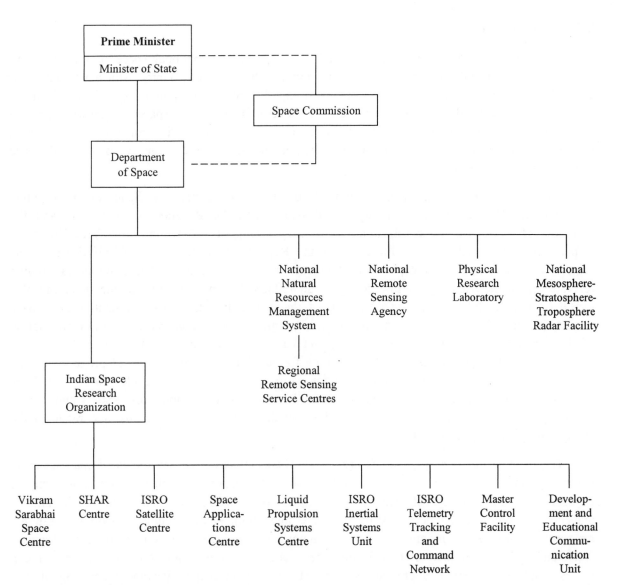

Figure 3.5 Organization of space activities in India

3.2.7 Indonesia

Indonesia has a well-established and fairly comprehensive national space programme. As the world's fourth and the Asian and Pacific region's third largest country with a population of about 200 million, abundant natural resources, a rapidly growing economy, expanding industry, a high priority accorded by the government to development of aerospace capabilities and an established institutional framework for matters relating to space and aeronautics, Indonesia is pursuing a space programme that can be categorized among the larger medium-sized space programmes in the region.

The national space agency, the National Council for Aeronautics and Space of the Republic of Indonesia (DEPANRI), is a high-powered body, with the President of Indonesia and the Minister of State for Research and Technology being its Chairman and Vice-Chairman respectively. The fact that the President of the Republic himself chairs DEPANRI shows the high priority attached to aerospace development in national affairs. Other ministers holding relevant portfolios, such as, for instance, Foreign Affairs, Defence and Communications, are also members of DEPANRI, while the Chairman of the National Institute of Aeronautics and Space (LAPAN), which is the principal national R and D organization responsible for the implementation of the aerospace programme in the country, serves as its Secretary. DEPANRI is the highest-level forum in the country responsible for policy formulation and coordination of programmes and activities relating to space and aeronautics for national development. Besides LAPAN, a large number of government ministries, departments and research organizations, universities and private sector agencies are involved in space activities, which is indicative of the size and range of the overall Indonesian space effort.

The national space programme is broad-based, encompassing several fields of space science, technology and applications. These include middle and upper atmosphere physics, ionospheric physics, HF radio communications, solar physics, remote sensing for resource and environmental management, pollution studies, ocean-atmosphere interaction studies, climate modelling, satellite meteorology and disaster monitoring, satellite communications and distance education, development and standardization of meteorological sounding rockets and development of ionosondes, among others. Indonesia operates a ground receiving station for the acquisition and processing of data from various remote sensing satellites, such as Landsat, SPOT, ERS-1 and JERS-1 at Pare Pare. There are also two local user terminals in Jakarta and Ambon linked to the international satellite-aided search and rescue COSPAS-SARSAT programme. Indonesia is also engaged in the development of its own aeronautical navigation satellite surveillance system for the twenty-first century. Further, it is seriously contemplating embarking upon the establishment of an international satellite launching station.

Indonesia was the third country in the world, after the United States and Canada, and the first among developing countries, to have its own operational domestic satellite communications system. It gained this distinction when the first of its Palapa communication satellites, built in the United States, was launched in 1976. Satellites offer the quickest, most practical and most cost-effective means of establishing communication links between the over 17,000 islands of the Indonesian archipelago. The successful operation of the Palapa system has amply demonstrated the usefulness and relevance of satellite communications to developing countries. Since 1976, three generations of Palapa satellites (Palapa-A, B and C) have been launched. The Palapa system is being used for a range of communication services, which include telephones, broadcasting, television, data communications and distance education. These satellites have also been used for communications by neighbouring ASEAN countries, including Malaysia, Philippines, Singapore and Thailand, through leasing of transponders. Further, a private consortium in Indonesia, which includes government-owned telecommunication agencies, plans to launch a satellite communications system called Indostar, comprising four direct broadcasting satellites being built by an American company. The first satellite of this system is scheduled for launch in 1997.

Indonesia attaches great importance to international and regional cooperation. It is thus contributing to various projects like IGBP, GlobeSAR and EC-ASEAN project. The Government of Indonesia also provides full local support for six trainees selected every year by ESCAP from developing countries for a four-month course in GIS applications at one of the best universities in this field in the country.

Further information on Indonesia's space programme may be obtained from the RESAP national focal point:

Professor Haryono Djojodihardjo
Chairman
LAPAN
Jalan Pemuda Persil #1
P.O. Box 1020/JAT
Jakarta
Indonesia
Fax: (62 21) 489-4815, 871-7715
E-mail: hdjojodihardjo@mail.lapan.go.in

National contact point for remote sensing/GIS/satellite-based positioning:

Mr Bambang Tejasukmana
Head, Data Bank Division
Remote Sensing Centre, LAPAN
Jalan Pemuda Persil #1
P.O. Box 1020/JAT
Jakarta
Indonesia
Fax: (62 21) 489-4815, 871-7715

I

General Information on National Space Activities

National space agency responsible for coordination of space activities	National Council for Aeronautics and Space of the Republic of Indonesia (DEPANRI).
Mandate	DEPANRI constitutes the highest national coordination and policy formulation forum in the country for national aerospace (air and space) development. The main task and function of DEPANRI is to assist the President in formulating national policy, regulations and utilization of aerospace for national development as well as in coordinating national space activities. The Chairman, Vice Chairman and Secretary of DEPANRI are the President of Indonesia, the Minister of State for Research and Technology and the Chairman of the National Institute of Aeronautics and Space respectively, while its members consist of the Minister of Foreign Affairs, the Minister of Defence, the Minister of Industry and Trade, the Minister of Communications, the Minister of Tourism, Post and Telecommunication, and the Minister of State for National Development Planning/the Chairman of the National Development Planning Board.
Organizational set-up	See figure 3.6.
Brief outline of the national space programme	Recent space activities in Indonesia can be grouped into space system and technology development, and research development and applications.

A. Space system and technology development

Space system and technology development include, but are not limited to, standardization of meteorological rockets, development and installation of ionosonde stations at various locations in the country, development of aeronautical navigation satellite system, and establishment of an international launch station. The standardization of meteorological rockets is aimed at building standardized meteorological rockets for the purposes of research dealing with middle and upper atmospheric physics. In support to the ionospheric research, four ionosonde stations have been installed and are operating. In furtherance of communications, navigation, surveillance and management in air traffic in the first decade of the twenty-first century, Indonesia is now initiating its own aeronautical navigation satellite system. Indonesia is also seriously considering embarking on the establishment and operation of an international launch station.

B. Research development and applications

A number of space activities dealing with research development and applications have been conducted. Data gathered by the ionosonde stations and other related data have been utilized for the frequency prediction of HF radio communication. Meteorological data and sea surface temperature have been used to build a climate prediction model for the whole country. Pollution measurement, ocean-atmosphere interaction study and development of methods for marine resources stock measurement have been undertaken using remote sensing data. Besides, remote sensing data and other relevant data have been used for development of rice crop prediction, flood monitoring and other natural hazards and disasters mitigation. Needless to say, since 1976 the area of space communication applications is the biggest business success of space technology applications in the country.

Organizations responsible for various facets of the national space programme

Remote sensing/GIS/satellite-based positioning

National Institute of Aeronautics and Space (LAPAN), National Coordination Agency for Surveys and Mapping, Agency for Assessment and Application of Technology (BPP Teknologi), Ministry of Public Works, Indonesian Institute of Sciences (LIPI), Bogor Agriculture University, University of Indonesia, Bandung Institute of Technology, Ministry of Forestry, Ministry of Mining and Energy, Ministry of Agriculture, Ministry of Transmigration, Ministry of Defence, Centre for Surveys and Mapping, Ministry of State for Environment, National Land Agency.

Satellite meteorology/disaster monitoring

LAPAN, Ministry of Forestry, BPP Teknologi, Meteorological and Geophysical Agency.

Satellite telecommunications/distance education

Ministry of Posts, Tourism and Telecommunications, Ministry of Education and Culture, Ministry of Internal Affairs, LAPAN, PT Telekominikasi, PT INDOSAT, PT Satelit Palapa Indonesia (PT Satelindo), PT Pasifik, Satelit Nusantara, PT Media Citra Indostar, Gedung Elektrindo.

Space sciences and technology applications

Space sciences and technology application activities are conducted by various institutions: BPP Teknologi, LIPI, universities and various private companies, LAPAN.

Legislation, government policy or directive related to integration of space applications in the development planning process

Indonesia has attached high priority to space technology in its national development programme. This has allowed many organizations to incorporate space technology applications in their activities. However, the priorities of the national development programme limit the focus of the space programme to space technology applications. Nevertheless, it is envisioned that the space programme will support, or even accelerate, the development of space as well as industrial activities to achieve self-sustainability.

Coordination and promotional activities

In the last two years, all national space-related institutions and agencies have been involved in formulating "The general policy on space development programme in the second long range development" (25 years ahead).

Status of space science and technology education

The University of Gadjah Mada conducts regular courses for undergraduate, graduate and postgraduate degrees in remote sensing applications. The Bandung Institute of Technology conducts regular courses for undergraduate and graduate degrees in space science and technology, and for postgraduate degrees in some frieds.

Other pertinent information or remarks

It is now planned to harmonize national policies to support involvement of the private sector in space activities.

Regional space applications programme network

Contact	Name	Agency
RESAP national focal point	Professor H. Wiryosumarto	National Institute of Aeronautics and Space
Regional Working Group on Remote Sensing, GIS and Satellite-based Positioning	Mr Bambang Tejasukmana	National Institute of Aeronautics and Space
Regional Working Group on Satellite Communication Applications	Mr Lukman Hutagalung	Directorate General of Post and Telecommunications, Department of Tourism, Post and Telecommunications
Regional Working Group on Meteorological Satellite Applications and Natural Hazards Monitoring	Mr Sujadi Hardjawinata	Meteorological and Geophysical Agency, Department of Communications
Regional Working Group on Space Sciences and Technology Applications	Mr Soewarto Hardhienata	National Institute of Aeronautics and Space

Respondent to questionnaires

Field	Agency	Name
Remote sensing, GIS and satellite-based positioning	National Institute of Aeronautics and Space	Mr Bambang Tejasukmana
Satellite communication applications	Ministry of Posts, Tourism and Telecommunications	Mr Danny Setiawan
Meteorological satellite applications and natural hazards monitoring	National Institute of Aeronautics and Space	Mr Agus Hidayat
	Meteorological and Geophysical Agency	Mr M. Nazamudin
Space sciences and technology applications	National Institute of Aeronautics and Space	Mr Ing Soewarto Hardhienata

I

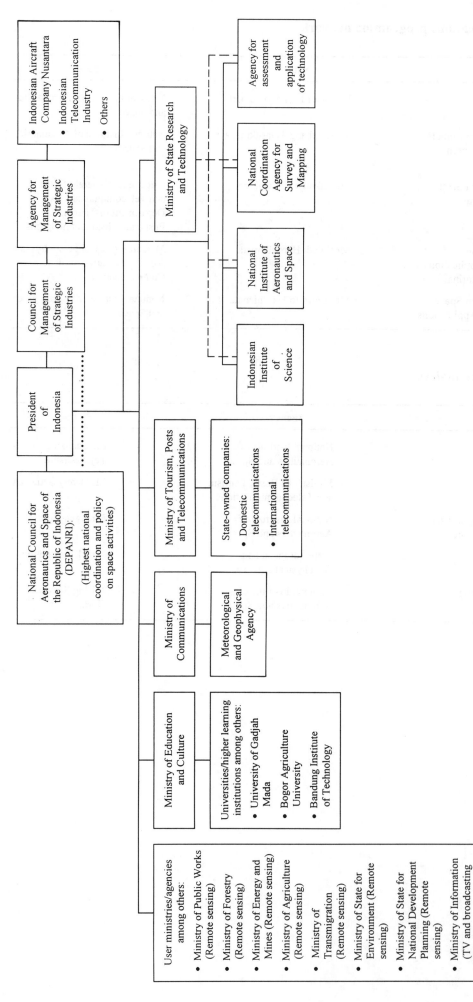

Figure 3.6 Organization of space activities in Indonesia

3.2.8 Islamic Republic of Iran

The national space programme of the Islamic Republic of Iran can be categorized as a medium-sized programme. The space programme is quite wide-ranging and is wider in scope than that of most developing countries. The Islamic Republic of Iran has substantial natural resources, especially huge reserves of oil, a fairly developed infrastructure and a large pool of qualified and trained personnel. Its space programme is thus expected to expand and diversify further.

The country established a ground station for receiving satellite (Landsat) remote sensing data at Mahdasht, about 50 km from Tehran, as far back as 1976, which would have been the earliest ground receiving station in the entire region had it been commissioned. Owing to a combination of various administrative and technical problems the station could not become operational. Efforts are under way to overcome these hurdles, suitably upgrade the station and make it fully operational.

In line with the broad recommendations of the Ministerial Conference on Space Applications for Development, held at Beijing in 1994, the Government of the Islamic Republic of Iran has set up a National Consultative Committee on Space Science and Technology Applications Affairs, headed by the Minister of Post, Telegraph and Telephone. The president of the Iranian Remote Sensing Centre (IRSC) serves as the secretary of this national consultative committee. IRSC, which is administratively under the Ministry of Post, Telegraph and Telephone (MPTT), is mandated to develop the country's remote sensing capabilities in all aspects of its technology and applications. It also functions as the main national agency responsible for overseeing and coordinating space activities in the country.

Remote sensing constitutes one of the principal components of the national space programme. Besides IRSC, which plays a lead role in space and remote sensing activities, approximately 50 agencies in the country are involved in remote sensing technology and its applications, including government ministries and organizations, universities and private sector agencies. Remote sensing, along with GIS, is being used by these entities in a variety of applications in agriculture, water resources management, geological mapping, mineral exploration, rangeland management, cartography, land-use planning and desertification. A national network for public information service has been set up to create computerized databases accessible to national users. IRSC has initiated two databases on earth resources satellite data and remote sensing education courses in the country. More databases are being planned by IRSC. Satellite-based positioning is being employed by several organizations in the country for accurate positioning of marine vessels, geodesy, navigation, rescue and relief operations and study of crust movements. Meteorological satellite data are also being extensively used, mainly by the national meteorological agency, the Islamic Republic of Iran Meteorological Organization, for weather forecasting, natural hazard monitoring, sea surface temperature studies and other related applications.

Seven universities in the country are conducting undergraduate and postgraduate courses covering different areas of space science and technology, including remote sensing, GIS, satellite meteorology, satellite communications, aeronautics and astronomy. There are, in addition, other scientific institutions and bodies engaged in space science research and education.

The Islamic Republic of Iran has a fairly extensive telecommunications infrastructure and network, which is being further extended to increase the number of available international channels by almost a factor of three and expand VSAT operations. The Telecommunication Company of Iran, affiliated to MPTT, is the main government agency involved in domestic communications. Another affiliate, the Iran Telecommunication Research Centre, carries out research and development projects in telecommunications at the national level. As part of the country's long-term telecommunications development programme, the government is planning to have a national satellite communications system based on three identical satellites, named ZOHREH.

Further information on the country's space programme may be obtained from the RESAP national focal point:

Mr Ahad Tavakoli
Chairman and Chief Executive Officer
Iranian Remote Sensing Centre
Ministry of Post, Telegraph and Telephone
P.O. Box 11365/6713
Tehran
Islamic Republic of Iran
Fax: (98 21) 206-4474
E-mail: itrc@www.dct.co.ir

General Information on National Space Activities

National space agency responsible for coordination of space activities	Iranian Remote Sensing Centre (IRSC), Ministry of Post, Telegraph and Telephone (MPTT).
Mandate	IRSC, as the main coordinator for remote sensing, GIS and GPS activities in the country, has interaction with over 300 government and private sector agencies and educational institutions and provides them with various services. IRSC has played a leading role developing remote sensing and GIS technologies in the country. It has thus been appointed to coordinate all tasks at national, regional and international levels, including tranferring technology to the country.
Organizational set-up	See figure 3.7.
Brief outline of the national space programme	The programme is broadly aimed at using and promoting space science and technology for the development of the country and the welfare of the people.
Organizations responsible for various facets of the national space programme	Ministry of Agriculture, Ministry of Industry, Islamic Republic of Iran Broadcasting Organization, Ministry of Foreign Affairs, Geological Survey of Iran, Forest and Range Organization, Bureau of Regional Planning, National Cartographic Centre, Ministry of Defence and Armed Forces, Ministry of Road and Transportation, Ministry of Culture and Education, Ministry of Post, Telegraph and Telephone.
Legislation, government policy or directive related to integration of space applications in the development planning process	The government directive which established the National Consultative Committee on Space Science and Technology Applications Affairs. The meetings of the Committee are regularly held at IRSC headquarters.
Coordination and promotional activities	The functions and duties of the National Consultative Committee relate to coordination and promotion of space science and technology applications in the country. As the national agency, IRSC is also involved in these activities.
Status of space science and technology education	At present more than seven universities and educational institutions in the country have academic and research programmes for space science and technology applications. These programmes include Earth observations, satellite meteorology, telecommunications, and space science and technology applications.

Regional space applications programme network

Contact	Name	Agency
RESAP national focal point	Mr Ahad Tavakoli	Iranian Remote Sensing Centre
Regional Working Group on Remote Sensing, GIS and Satellite-based Positioning	Mr Ahmad Mohammadpour	Iranian Remote Sensing Centre
Regional Working Group on Satellite Communication Applications	Mr Mohammad Hakak	Iranian Telecommunication Research Centre
Regional Working Group on Meteorological Satellite Applications and Natural Hazards Monitoring	Ms Mehrnaz Bijanzadeh	Islamic Republic of Iran Meteorological Organization
Regional Working Group on Space Sciences and Technology Applications	Mr Mohammad-Hasan Entezari	Data Communication of Iran

Respondent to questionnaires

Field	Agency	Name
Remote sensing, GIS and satellite-based positioning	Iranian Remote Sensing Centre	Mr Ahmad Mohammadpour
Satellite communication applications	Iran Telecommunication Research Centre	Mr Abdol Mohammad Darab
Meteorological satellite applications and natural hazards monitoring	Islamic Republic of Iran Meteorological Organization	Ms Mehrnaz Bijanzadeh
Space sciences and technology applications	Telecommunication Company of Iran	Mr Mohammad Reza Golabchian

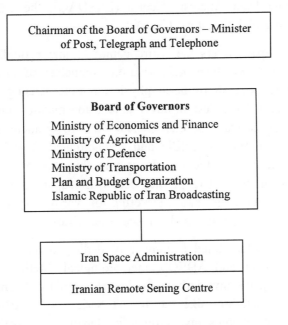

Figure 3.7 **Organization of space activities in the Islamic Republic of Iran**

3.2.9 Japan

Japan is one of the few countries in Asia that is highly developed and industrialized. As one of the world's strongest economies and largest producers of sophisticated technology, Japan is pursuing an advanced and comprehensive space programme encompassing a wide range of fields related to space science, technology and applications.

The overall responsibility for coordinating the national space programme in Japan has been assigned to the Space Activities Commission (SAC), with the Space Policy Division of the Science and Technology Agency (STA) acting as its secretariat. SAC integrates and unifies the space-related activities of government agencies, and submits analyses and opinions regarding policies, coordination, budget estimates and other relevant matters to the Prime Minister's office for policy formulation. STA is involved in drawing up the actual plans and programmes relating to space, in accordance with the policy made at the higher levels, for implementation through the National Space Development Agency of Japan (NASDA), the Institute of Space and Astronautical Science (ISAS) and the National Aerospace Laboratory (NAL). As the secretariat for SAC, STA is responsible for coordination and liaison among the various government agencies. It also annually reviews the space development programme formulated in line with the fundamental national policy.

NASDA, which consists of several constituent departments specializing in various specific sectors and activities, such as launch vehicles, propulsion systems, satellites and ground stations, functions as the main agency in the country responsible for the implementation of the national space programme, especially the development of the space segment and associated ground facilities. Besides NASDA, there are other organizations, such as NAL, ISAS of the Ministry of Education, Science and Culture, the Meteorological Satellite Centre of the Japan Meteorological Agency, and the Communications Research Laboratory of the Ministry of Posts and Telecommunications that are actively involved in space-related activities.

The Ministry of International Trade and Industry (MITI) is also involved in space activities related to those sectors, such as environment, trade and energy, that fall within its purview. Some of the organizations under MITI involved in space activities are the Space Industry Division, the Earth Remote Sensing Data Analysis Centre (ERSDAC), the Geological Survey of Japan (GSJ) and the Electrotechnical Laboratory (ETL).

The space effort in Japan covers the entire gamut of major fields related to space science, technology and applications, which includes various branches of astronomy and astrophysics, magnetospheric, ionospheric and atmospheric physics, remote sensing, satellite meteorology, life sciences, materials science, and the design, development and manufacture of satellites, subsystems, sensors, launch vehicles, space stations, etc. Besides the government agencies mentioned above, there are a large number of other scientific and technological R and D institutions as well as universities in both the public and the private sectors that are involved in different aspects of space research and applications. The private sector in Japan is extensively involved in the research, development, manufacture and marketing of various types of equipment, systems, instruments, components and materials used for space activities.

Japan has designed and fabricated a number of dedicated remote sensing satellites such as MOS-1a and MOS-1b, JERS-1 and ADEOS, meteorological satellites such as the GMS series, and other satellites using Japan's own indigenous launchers of the N and H (H-I and H-II) series. Operational communication satellites like the JCSAT series, SUPERBIRD, CS/N-Star and BSat, owned by private sector companies and groups in Japan, have been built by United States manufacturers and launched by United States or European launch service providers. The country has ground receiving stations at Hatoyama and Kumamoto for handling data from its indigenously developed remote sensing satellites such as MOS, JERS and ADEOS as well as from Landsat,

SPOT and ERS. Other more advanced Earth observation satellites such as TRMM, ADEOS-2 and ALOS are under design and development and are planned for launch between 1997 and 2002.

Japan has developed and launched some scientific satellites/spacecraft for various purposes, such as MUSES-A to orbit the moon, EXOS-D to detect X-rays from a supernova, Sakigake to study Halley's Comet and GEOTAIL to study the magnetosphere.

In recent years, Japan has been laying greater emphasis on international cooperation in the domain of space, specially with the countries of the ESCAP region. Japan is also the single biggest donor country for space-related programmes and activities, especially those conducted under the aegis of RESAP. Under a joint RESAP-NASDA research programme, Japan was to provide substantial funding support and ADEOS data for 20 principal investigators for 20 selected research studies based on ADEOS data in 15 developing countries of the region. Japan also cooperates with RESAP in organizing annual seminar-cum-training programmes on Earth observation technology applications. Japan also has growing cooperation on the wider international plane with other countries and groups such as the United States, Canada, Europe (through ESA) and individual European countries, and the Russian Federation. For instance, the TRRM is being jointly developed with in the United States NASA, while ADEOS has on-board sensors provided by NASA and JPL and France's CNES. Japanese astronauts flew on United States space shuttle missions in 1992 and 1994. Further, Japan will be a major partner with the United States, Canada, Europe and the Russian Federation in building and maintaining a permanent international space station planned to start operating fully by the year 2002.

Further information on Japan's space programme may be obtained from the RESAP national focal point:

Mr Eiichi Kawahara
First Secretary and Deputy Permanent Representative of Japan to ESCAP
Embassy of Japan
Bangkok, Thailand

J

General Information on National Space Activities

National space agency responsible for coordination of space activities	Space Activities Commission (SAC) and the Science and Technology Agency (STA).
Mandate	The purpose of SAC is to unify space activities of various government agencies and actively promote them. SAC plans, deliberates and decides on the matters below and submits its opinions to the Prime Minister. The Prime Minister should respect the opinions of SAC on important policies regarding space activities, important matters related to overall coordination of space-related work among government agencies, estimates of space activity expenses of government agencies, matters related to cultivation and training of space researchers and engineers (excluding instruction and research at universities or colleges) and other important matters concerning space activities.
	STA plans programmes and promotes basic space-related policy and overall coordination of space activities among governmental agencies, and conducts research and development activities through the National Aerospace Laboratory (NAL) and National Space Development Agency of Japan (NASDA). As the secretariat of SAC, STA conducts liaison and coordination activities among governmental agencies.
Organizational set-up	See figures 3.8 and 3.9.
Brief outline of the national space programme	Japan's space development is conducted in accordance with the long-term guidelines stated in the Fundamental Policy of Japan's Space Development revised by SAC in January 1996. A brief outline of the Fundamental Policy is as follows:

(1) Full-scale utilization towards the twenty-first century:

- Contribution to the preservation of global environment by Earth Observing Satellite series

- Promotion of space environment utilization by Japanese Experiment Module of the International Space Station
- Development of high-level utilization of satellites by Mission Demonstration Satellite series etc. (high-speed satellite communication technology, digital satellite broadcasting, navigation technology etc.): Realization of economical space activities (H-IIA rocket, etc.)

(2) Creative space development towards new areas:

- Challenge to the creative space development such as unmanned lunar exploration and HOPE-X etc.
- Active promotion of international cooperation in the area of global earth observation system

(3) Space development properly reflecting social requirements:

- Timely evaluation of the progress and the outcome of space development
- Steady expansion of the governmental budget and efforts to eliminate unnecessary expenses
- Consideration of the healthy development of space industry
- Consideration of the preservation of the space environment
- Strengthening public relations to deepen the understanding of citizens

Organizations responsible for various facets of the national space programme	National Aerospace Laboratory; National Space Development Agency of Japan; Institute of Space and Astronautical Science (ISAS), Ministry of Education, Science and Culture; Meteorological Satellite Centre, Japan Meteorological Agency; Communications Research Laboratory, Ministry of Posts and Telecommunications.
Legislation, government policy or directive related to integration of space applications in the development planning process	None.
Coordination and promotional activities	SAC annually revises the Space Development Programme which formulates specific projects currently required for Japan, based on progress of research and development, in line with the Fundamental Policy.
Status of space science and technology education	It is stated in the Fundamental Policy that more emphasis should be placed on the undergraduate and postgraduate educational systems; training of researchers and engineers in space-related agencies should be strengthened; and researchers and engineers in the Asian and Pacific region should be assisted by accepting them in space-related agencies.

Regional space applications programme network

Contact	Name	Agency
RESAP national focal point	Permanent Representative to ESCAP	Embassy of Japan
Regional Working Group on Remote Sensing, GIS and Satellite-based Positioning	Mr Akira Noie	Science and Technology Agency
Regional Working Group on Satellite Communication Applications	Mr Masakuni Esaki	Ministry of Posts and Telecommunication
Regional Working Group on Meteorological Satellite Applications and Natural Hazards Monitoring	Mr Masaro Saiki	Japan Meteorological Agency
Regional Working Group on Space Sciences and Technology Applications	Mr Akira Noie	Science and Technology Agency

Respondent to questionnaires

Field	Agency	Name
Remote sensing, GIS and satellite-based positioning	National Space Development Agency	Mr Sohsuke Gotoh
	Geological Survey of Japan, Ministry of International Trade and Industry	Mr Koji Takimoto
	Earth Remote Sensing Data Analysis Centre, Ministry of International Trade and Industry	Mr Koji Takimoto
Satellite communication applications	Telecommunication Advancement Organization	Mr Takamasa Suzuki
Meteorological satellite applications and natural hazards monitoring	Japan Meteorological Agency	Mr Tetsu Hiraki
Space sciences and technology applications	Electrotechnical Laboratory, Ministry of International Trade and Industry	Mr Kazuo Machida
	Institute of Space and Astronautical Science	Mr Kazuo Anazawa

J

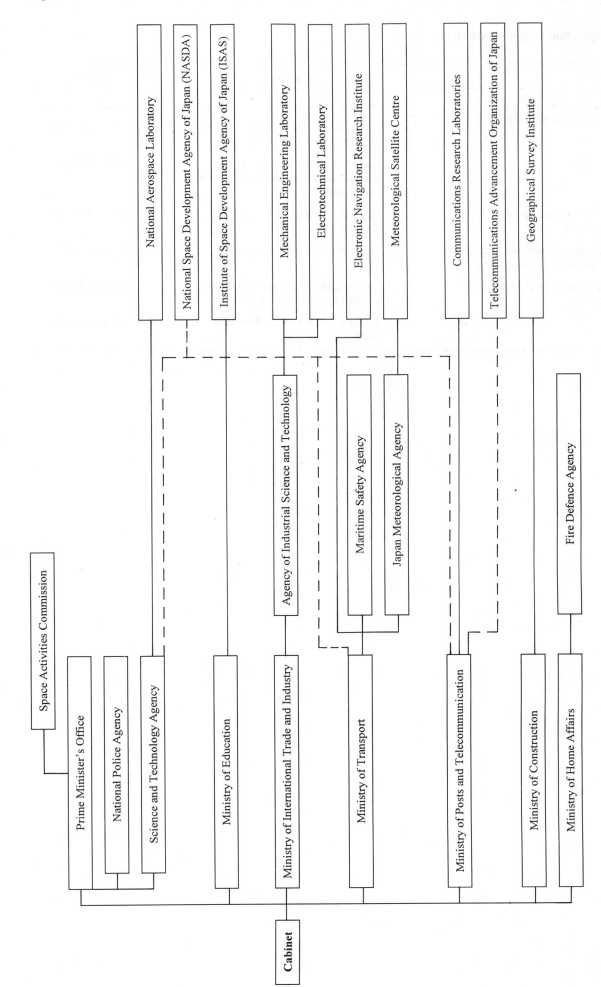

Figure 3.8 Organization of space activities in Japan

J

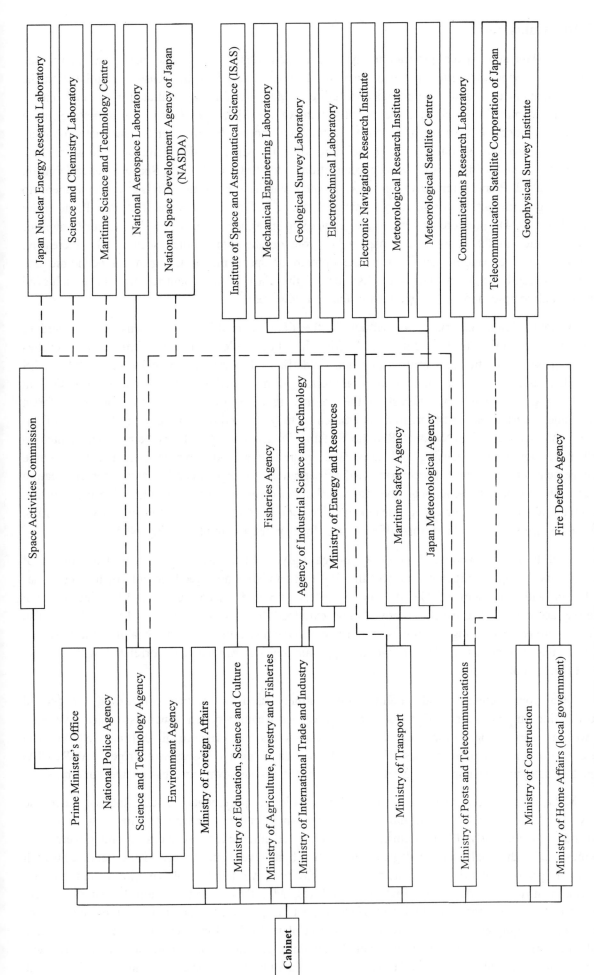

Figure 3.9 Linkages between NASDA and affiliated Japanese organizations

3.2.10 Malaysia

The Malaysian space programme has developed steadily over the past few years (see figure 3.10). Space activities in the country, until about a decade ago, focused essentially on the use of communication and meteorological satellites. Malaysia formally initiated a national remote sensing programme in 1989 with the establishment of the Malaysian Centre for Remote Sensing (MACRES) under the Ministry of Science, Technology and the Environment. The Meteorological Services are also administratively under the same Ministry, while satellite communications are dealt with by the Ministry of Energy, Telecommunications and Post. In 1992, a Space Science Division (initially called Planetarium Division) was created within the Prime Minister's Department to assist the government in formulating a national space policy, manage the National Planetarium (opened in 1994), promote research and education in space science in the country, monitor space activities in the country, and play a key role on the international scene. The increasing investments and expanding activities in the domain of space mirror the growing economic and financial strength of Malaysia, which is one of the East Asian countries that have achieved phenomenal socio-economic progress over the last few years.

Although Malaysia was a late entrant in the field of satellite remote sensing compared to many other countries in the region, MACRES has been able to develop impressive facilities and expertise in this field, as well as in allied fields like GIS over the last eight years. MACRES and other user agencies in the country are engaged in applying remote sensing and GIS in a wide range of studies of interest to the country relating, for instance, to monitoring forest cover and forest areas, soil erosion assessment, study of sedimentation, and detection of oil spills. A comprehensive National Remote Sensing Programme (NRSP) is being implemented to further develop remote sensing and associated technologies, and promote their use in resource management, environmental momitoring and preservation, strategic planning and other application areas. NRSP emphasizes, among other aspects, human resources development, active participation in international programmes, and greater involvement of the private sector in NRSP activities. MACRES plays a key role in this effort as the focal point for implementing NRSP.

Satellite-based positioning is being increasingly used in Malaysia. The Department of Surveys and Mapping initiated a project in 1989 to establish a new geodetic control for peninsular Malaysia with the help of GPS. The project completed successfully in 1993 led to high relative accuracy. The work is now being extended to East Malaysia. MACRES has also initiated projects involving integration of GPS with GIS and remote sensing enabling the positions of test sites and points on the ground to be accurately determined and correlated with satellite images in a short time.

Data from meteorological satellites are routinely used in weather forecasting for general and aviation purposes, monitoring potentially disastrous weather systems such as tropical cyclones, as well as marine applications such as shipping, fisheries and offshore oil drilling. NOAA data were also extensively used in phase II of the Agroclimatic Impact Assessment Project (AGROCIA) aimed at studying the impact of weather on rice yield in rice-growing areas of the country.

Malaysia has been meeting its telecommunication requirements through Intelsat since the early 1970s. Later, it additionally leased transponders on Palapa to meet the growing demand. By the early 1990s, demand for satellite communication services among government, business and individual users, had grown to such a high level, due mainly to the continuing high economic growth, that it was considered feasible to have a national satellite communication system, named MEASAT. The first two satellites, MEASAT-1 and 2, built by Hughes, were launched by Arianespace in 1995 and 1996 respectively, and are operated by a private service provider. To complement the MEASATs, Malaysia plans to develop microsatellites in cooperation with the University of Surrey in the United Kingdom, as part of its small satellite development programme. In addition, Malaysia has active collaboration in respect of small satellites with relevant agencies in other countries, notably Brazil, India and South Africa. The Malaysian space programme is expected to expand substantially in the near future.

Further information on Malaysia's space programme may be obtained from the RESAP national focal point:

Mr Nik Nasruddin Mahmood
Director
Malaysian Centre for Remote Sensing (MACRES)
Ministry of Science, Technology and the Environment
Letter Box 208, Lot CB 100, 5th Floor
City Square Centre, Jalan Tun Razak
50400 Kuala Lumpur
Malaysia
Fax: (09 3) 264-5646
E-mail: macres@macres.sains.my

Regional space applications programme network

Contact	Name	Agency
RESAP national focal point	Mr Nik N. Mahmood	Malaysian Centre for Remote Sensing
Regional Working Group on Remote Sensing, GIS and Satellite-based Positioning	Mr Nik N. Mahmood	Malaysian Centre for Remote Sensing
Regional Working Group on Satellite Communication Applications	Mr Mohd Aris Bernawi	Department of Telecommunications
Regional Working Group on Meteorological Satellite Applications and Natural Hazards Monitoring	Mr Lim Joo Tick	Malaysian Meteorological Service
Regional Working Group on Space Sciences and Technology Applications	Professor Mazlan Othman	Planetarium Negara

Respondent to questionnaires

Field	Agency	Name
Remote sensing, GIS and satellite-based positioning	Malaysian Centre for Remote Sensing	Mr Darus Ahmad
Satellite communication applications	–	–
Meteorological satellite applications and natural hazards monitoring	Malaysian Meteorological Service	Mr Alui Bahari
Space sciences and technology applications	–	–

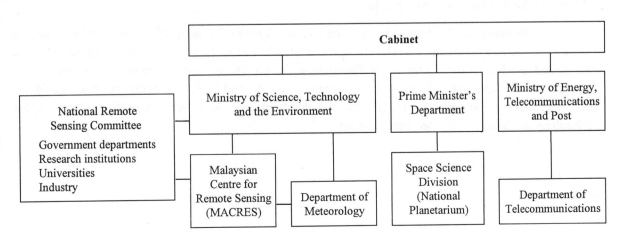

Figure 3.10 Organization of space activities in Malaysia

3.2.11 Mongolia

Mongolia has a small space programme, which is slowly developing. Remote sensing and GIS activities, which comprise the bulk of its space programme, are coordinated by the National Remote Sensing Centre (NRSC), an agency created in 1987 directly under the Ministry for Nature and the Environment.

NRSC is mandated to formulate policies on integrating remote sensing with national development, promote information exchange at national and international levels and provide technical training in this field. In recent years, remote sensing applications have grown from pilot experimental studies to operational methodologies. For example, NRSC has developed a technique to generate drought and snow maps from NOAA AVHRR data every 10 days and fire risk maps from the same source data every spring season.

Other application projects on remote sensing and GIS which NRSC has undertaken include thematic mapping, natural resource inventory and monitoring, natural hazards assessment, and climate change. Remote sensing and GIS technology has a wide user base with 15 departments of different ministries, 20 research institutes and 15 universities and colleges doing work and research in this field. Non-governmental organizations and private companies have recently engaged in remote sensing-related activities such as the Mongolian Association for Protection of Nature and Environment, Monmap Co., Ltd., and Orchlon Co., Ltd.

Space activities relating to satellite communication, which include television broadcasting, voice telephony and non-voice services, are coordinated by the Ministry of Infrastructure and carried out by the International Telecommunications and Cooperation Department of the Mongolian Telecommunications Company. Mongolia has an earth satellite ground station at Ulaanbaatar, installed at a cost of $12.6 million, which receives Intelsat, AsiaSat and Intersputnik signals.

The only organization involved in satellite meteorology is the Hydrometeorological Department of Mongolia. The agency maintains a network of weather forecasting bureaux in the countryside and provides them with cloud, snow cover, vegetation and soil moisture maps on a regular basis. It also maintains a meteorological satellite data reception facility which is capable of receiving and processing data from the NOAA, Meteor and GMS satellites.

In 1989, Mongolia adopted a policy of opening to the outside world and has since established extensive cooperative relations with other countries and international organizations. In the field of space technology, Mongolia has established cooperative relations with, among others, JICA to conduct a forest resource management study, Centre for Environmental Remote Sensing of Chiba University in Japan to conduct grassland and land cover assessment studies and the National Space Development Agency of Japan to organize an international seminar on space informatics for sustainable development.

The Government of Mongolia, in response to Agenda 21 adopted at the United Nations Conference on Environment and Development in 1992, formulated a Mongolian Agenda 21 and established the Mongolian Council for Sustainable Development. Both emphasize the importance of space technology applications, especially remote sensing and GIS, in making informed decisions regarding the environment and sustainable development. Thus, it sees increasing use of space technology applications in the years to come.

Further information on Mongolia's space programme may be obtained from the RESAP national focal point:

Mr Sodov Khudulmur
Director
National Remote Sensing Centre
Ministry for Nature and the Environment
Government Building No. 3, Baga Toirun-44
Khudaldaany str-5
Ulaanbaatar-11
Mongolia
Tel: (976 1) 312-269
Fax: (976 1) 329-968, 323-401

General Information on National Space Activities

National space agency responsible for coordination of space activities	National Remote Sensing Centre of Mongolia (NRSC), Ministry of Nature and Environment.
Mandate	To develop space programmes and policy/strategy; to expand space applications; to coordinate space programmes, activities and projects; and to promote space technology and facilities in agencies involved in remote sensing agencies
Organizational set-up	See figure 3.11.
Brief outline of the national space programme	Natural resources management, natural disaster monitoring, development of environmental databases based on remote sensing and GIS
Organizations responsible for various facets of the national space programme	Institute of Meteorology, Ministry of Agriculture, Ministry of Nature and Environment, Ministry of Geology and Industries, and other government organizations.
Legislation, government policy or directive related to integration of space applications in the development planning process	None.
Coordination and promotional activities	
Status of space science and technology education	Educational programmes are arranged in the following fields: image processing, satellite-based positioning, GPS, GIS and satellite meteorology.

M

Regional space applications programme network

Contact	Name	Agency
RESAP national focal point	Mr Sodov Khudulmur	National Remote Sensing Centre of Mongolia
Regional Working Group on Remote Sensing, GIS and Satellite-based Positioning	Mr Sodov Khudulmur	National Remote Sensing Centre of Mongolia
Regional Working Group on Satellite Communication Applications	Mr Sanjaa Ganbaatar	Telecommunication Authority of Mongolia
Regional Working Group on Meteorological Satellite Applications and Natural Hazards Monitoring	Ms M. Erdene-tuya	Hydrometeorological Department
Regional Working Group on Space Sciences and Technology Applications	Mr M. Ganzorig	Mongolian Academy of Sciences

Respondent to questionnaires

Field	Agency	Name
Remote sensing, GIS and satellite-based positioning	National Remote Sensing Centre of Mongolia	Mr M. Badarch
Satellite communication applications	Telecommunication Authority of Mongolia	Mr Sanjaa Ganbaatar
Meteorological satellite applications and natural hazards monitoring	National Remote Sensing Centre of Mongolia	Mr M. Badarch
Space sciences and technology applications	National Remote Sensing Centre of Mongolia	Mr M. Badarch

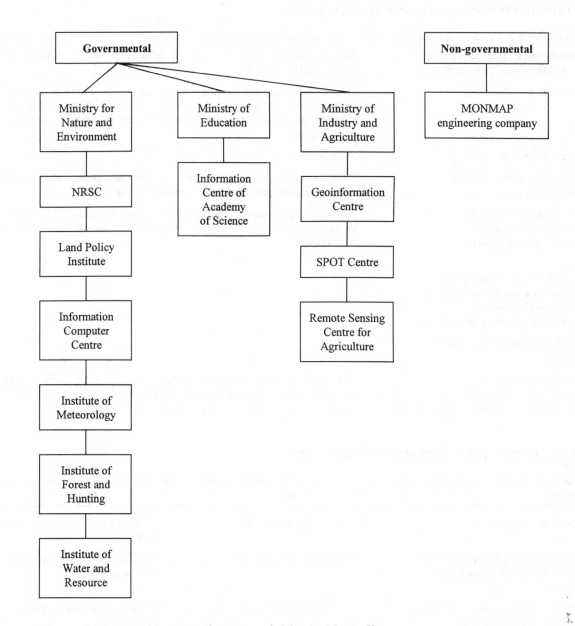

Figure 3.11 Organization of space activities in Mongolia

3.2.12 Myanmar

Myanmar, a medium-sized developing country intermediate between South and South-East Asia, is involved in a relatively modest space effort. The country does not, therefore, have a national space agency. National space activities in Myanmar are focused on the applications of remote sensing, GIS, satellite meteorology and satellite communications, as would be expected of a typical developing country confronted with the usual problems of resource constraints, inadequate technical skills and poorly developed infrastructure. Myanmar is also experiencing various environmental and ecological problems such as deforestation, erosion and floods, and satellite remote sensing data and GIS have been, and are being, used to address such problems.

There is no particular department or organization in Myanmar that serves as the country's designated national centre for remote sensing and GIS, although several departments and organizations are involved in remote sensing and GIS applications in their respective fields of interest. The Department of Meteorology and Hydrology has used Landsat data for flood inundation and related environmental studies within the framework of a WMO/UNDP project. This Department in a way acts as the focal or coordinating agency in Myanmar for space/ remote sensing activities in the context of international or regional programmes like RESAP, as its personnel have been interacting with ESCAP on behalf of the country in activities organized under the aegis of RESAP. This Department also operates a ground receiving station for reception of visible and near infrared from NOAA and GMS satellites. These are utilized for weather forecasting, especially for detection of vertically developing cumulo-nimbus clouds. There is also a radar station which is used to detect storms near the Arakan coast.

The Department of Forestry implemented the UNDP-funded National Forest Management and Inventory Project and National Forest Survey Project which had made extensive use of Landsat data and GIS software. The projects were completed in 1992-1993. The Department of Forestry has Arc Info and IDRISI, some Landsat data and a few trained specialists. Support was received from the International Centre for Integrated Mountain Development to upgrade their hardware and software and to provide training for their personnel. Several geographic databases and land-use maps were produced on a pilot scale using GIS. The department recently acquired three units of GPS receivers which they use for georeferencing satellite imagery and routine surveying.

M

Some other departments also have satellite remote sensing data and GIS facilities, but possess rather limited experience and skills in their application. Some departments have also shown interest in using satellite-based positioning, such as GPS.

Myanmar Posts and Telecommunications, under the Ministry of Communications, Posts and Telegraph, which is responsible for satellite communications in the country, upgraded inter- national communications facilities in the country by opening two Grade A earth stations in 1994. Myanmar uses the Intelsat, AsiaSat and Thaicom satellite systems to meet its satellite communi- cation requirements, including television broadcasting and telephone, telex and fax services. Satellite communications are also being used for distance education. Lectures from the Distance Education University are broadcast throughout the country, including remote areas, through satellite relay stations.

While space technology has been used in various application areas in Myanmar, there is a need to intensify efforts to use this versatile technology for various nation-building activities through effective coordination and cooperation between different relevant departments and organizations within the country.

An organization chart relating to space activities in Myanmar is given in figure 3.12.

Further information on Myanmar's space programme may be obtained from the RESAP national focal point:

Mr Htay Aung
Director-General
Department of Meteorology and Hydrology
Kaha-Aye Pagoda Road
Yangon
The Union of Myanmar
Fax: (95 1) 65944

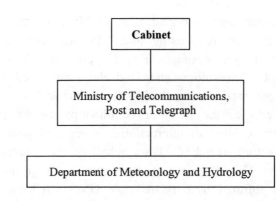

Figure 3.12 Organization of space activities in Myanmar

3.2.13 Nepal

Nepal is one of the smaller countries of the South Asian sub-continent. It is a landlocked country and much of its area is covered by the world's biggest mountain range, the Himalayas. Nepal would be included among the less developed countries of the region facing typical problems such as a low per capita income, financial constraints, low literacy level, shortage of technically qualified and skilled manpower, poor health care facilities and standards, increasing pressure on available land and natural resources, and weak infrastructure. The country is also afflicted by various environmental problems in the shape of erosion, landslides, floods, deforestation and related ecological imbalances. Owing to the steep Himalayan relief and fragile geology resulting from tectonic movement, Nepal is particularly vulnerable to water-induced disasters, especially in the monsoon season from June to September.

The space effort in Nepal is, accordingly, mainly focused on the use of remote sensing and GIS for resource and environmental surveying, and management and disaster monitoring. In 1981, the government established the National Remote Sensing Centre (NRSC), through the joint cooperation between the Government of Nepal and USAID, with the aim of bringing together multidisciplinary scientists to work on remotely sensed data to generate useful information for national development. The Centre served as the focal point for all remote sensing and GIS activities and regularly conducted seminars and workshops.

In 1989, however, the Centre was merged with the Forest Survey and Statistics Division of the Forest Research and Survey Centre and renamed the Remote Sensing Section, and it focused its work on forest resource inventory and monitoring land-use changes. In the absence of a national space agency, this Centre now acts as the focal or coordinating agency for space and remote sensing activities in the country. This Centre and several other agencies such as the Department of Geology and Mines, Nepal Agricultural Research Council and the Department of Hydrology and Meteorology are applying remote sensing data (Landsat and IRS) and GIS for various projects and studies relating to forest mapping, geological mapping, mining, agriculture, landcover and landuse classification, land resources assessment and planning, water resources assessment, watershed management, etc.

GIS is used by various government agencies and NGOs like the Integrated Centre for Mountain Development (ICIMOD), Department of Geography of Tribhuvan University, Survey Department, and a few private agencies. The Geography Department of Tribhuvan University has been recently collaborating with ICIMOD in GIS training and education.

In view of the advantages of satellite-based positioning, the Department of Survey has been using their six units of SR 200 GPS receivers for cadastral surveys, geodetic surveys and topographic mapping.

An APT receiving station for meteorological satellites has been functioning at Tribhuvan International Airport under the Department of Hydrology and Meteorology since 1975. It receives visible and infra-red images from NOAA satellites daily and from Meteor satellites occasionally. It was upgraded in 1990 to receive hourly images from Japan's GMS satellites. Although Nepal is at the western fringe in GMS images, the images are useful in summer when most of the cloud cover originates in the Bay of Bengal and further east. The station also receives pictures from Indian satellites. The images are used for routine weather forecasting, especially for civil aviation and tourist mountain expeditions.

The agency responsible for satellite communication in the country is the Nepal Telecommunication Corporation, which acts as the operating agency for the Intelsat system. The major function of the Corporation is to operate domestic and international telephone, telex and fax services and leased data circuits. It also provides television transmission services to Nepal television. The Corporation started its activities by establishing a standard B Sagarmatha Earth

N

station at Kathmandu in 1982. The Sagarmatha station was upgraded to standard A in 1994, with funding of $3.5 million from the World Bank, by the addition of IDR equipment, low rate encoding and cross-connect equipment.

Further information on Nepal's space programme may be obtained from the RESAP national focal point:

Mr Rajendra B. Joshi
Executive Director
Forest Survey and Survey Centre
P.O. Box 3339
Babar Mahal
Kathmandu
Nepal
Fax: (977 1) 220-159, 226-944
E-mail: foresc@wlink.com.up

General Information on National Space Activities

National space agency responsible for coordination of space activities	Forest Research and Survey Centre.
Mandate	To apply remote sensing and GIS to forestry and coordinate with various related agencies regarding space applications.
Organizational set-up	See figure 3.13.
Brief outline of the national space programme	Space activities in Nepal are confined to (a) application of satellite remote sensing data in natural resources surveying covering forestry, agriculture, geology and mines, weather forecasting and hydrological studies, and (b) telecommunications and television broadcasting.
Organizations responsible for various facets of the national space programme	National-level forest survey activities are carried out by the Forest Research and Survey Centre. Hydrological and meteorological data collection and weather forecasting are carried out by the Department of Hydrology and Meteorology. Provision of telecommunication services and training is the responsibility of the Nepal Telecommunication Corporation.
Legislation, government policy or directive related to integration of space applications in the development planning process	None.
Coordination and promotional activities	The networking of programmes/institutions within the country is under discussion.
Status of space science and technology education	The Department of Geography at Tribhuvan University conducts short courses in GIS.

Regional space applications programme network

Contact	Name	Agency
RESAP national focal point	Mr Rajendra B. Joshi	Forest Research and Survey Centre
Regional Working Group on Remote Sensing, GIS and Satellite-based Positioning	Mr Rajendra B. Joshi	Forest Research and Survey Centre
Regional Working Group on Satellite Communication Applications	–	–
Regional Working Group on Meteorological Satellite Applications and Natural Hazards Monitoring	Mr Kiran Shankar Yogacharya	Department of Hydrology and Meteorology
Regional Working Group on Space Sciences and Technology Applications	Mr Keshar M. Bajracharya	Royal Nepal Academy of Science and Technology

Respondent to questionnaires

Field	Agency	Name
Remote sensing, GIS and satellite-based positioning	Forest Research and Survey Centre	Mr Rajendra B. Joshi
Satellite communication applications	Nepal Telecommunications Corporation	Mr B.R. Kanel
Meteorological satellite applications and natural hazards monitoring	Department of Hydrology and Meteorology	Mr Kiran Shankar Yogacharya
Space sciences and technology applications	–	–

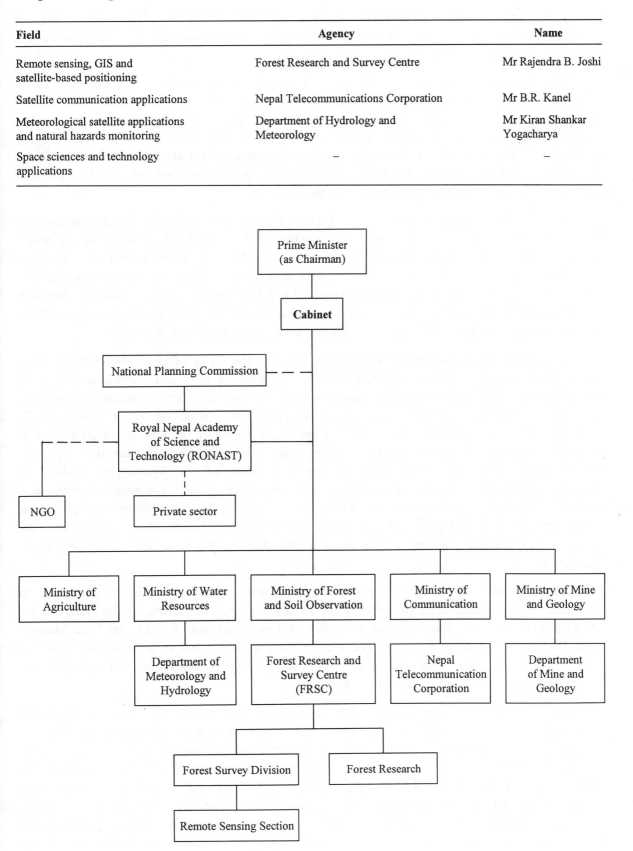

Figure 3.13 Organization of space activities in Nepal

N

3.2.14 New Zealand

New Zealand is the southernmost country in the ESCAP region. This small country of about 3.5 million people would be categorized among the comparatively more prosperous and developed countries of the region.

While New Zealand has a long experience in carrying out space-related activities, particularly in remote sensing, GIS and satellite-based positioning, it does not have a national space agency or a formal national government-supported space programme. However, several government, Crown Research and private sector organizations in the country are involved in various space applications. The main agency that is involved in most space applications activities is Landcare Research (NZ), Ltd., a Crown Research institute. This agency therefore usually plays the lead role in space-related matters and activities in the country.

The overall space effort is not very extensive given the small size of the country and its requirements. However, as a developed country with a good infrastructure and a sound education and R and D base, New Zealand has appreciable expertise in several areas of space technology applications. The space effort covers several fields including remote sensing, GIS, satellite meteorology, satellite-based positioning and satellite communications. Landcare Research is well-equipped with facilities for processing and analysis of data from remote sensing satellites, mainly Landsat, SPOT and NOAA, aerial photographs and conventional data sets.

In addition, three government universities, namely, the universities of Auckland, Waikato and Wellington, and a number of private agencies are involved in remote sensing activities. In fact, the role of the private sector in remote sensing and satellite communications is growing in the country. Remote sensing technology is being used for a wide variety of applications in New Zealand. For instance, SPOT and TM are being employed in land degradation studies, landslide mapping, forest cover mapping, vegetation monitoring and biomass estimation, especially in hilly terrain, while SAR data are being applied in geological and hydrological studies. In research studies relating to the application of both optical and microwave remote sensing data, there is a great deal of emphasis on development of algorithms, improvements in processing techniques and methodologies, enhancement in the accuracy of automated classification, and so on. Most research and applications work in the field of remote sensing in the country is user-driven.

NOAA AVHRR HRPT data are received from ground stations at Wellington and at Lower Hutt and are being utilized in the country for routine weather forecasting by the New Zealand Meteorological Service. Also engaged in satellite meteorological activities are the Land Management Division of Landcare Research, National Institute of Water and Atmosphere Research and some private companies. These data are also used for non-meteorological purposes, like sea surface temperature estimation and NDVI monitoring.

GIS is extensively employed by many government and private agencies as a necessary component in projects or studies relating to land resources mapping, grassland monitoring, soil classification, crop monitoring, forest inventories, and environmental degradation. The Land Spatial Information Group of Landcare and the Department of Survey and Lands Information are the primary agencies responsible for the maintenance of two national data bases, namely, New Zealand Land Resource Database and New Zealand Soils Database. Various other national resource databases are being developed by different concerned agencies. Similarly, GPS is also being increasingly used in the country for various projects, specially for crustal deformation and plate tectonic studies.

Government and private agencies in New Zealand have a fair amount of interaction and liaison with the other small island countries in the Pacific subregion, and this is expected to increase with the passage of time.

Further information on New Zealand's space programme may be obtained from the RESAP national focal point:

Mr David Pairman
Landcare Research
P.O. Box 69
Lincoln 8152
New Zealand
Tel: (64 3) 325-6700
Fax: (64 3) 325-2418

(Information on New Zealand's involvement in the four fields of space technology applications covered in this inventory could not be obtained in time for the production of this document.)

N

3.2.15 Pakistan

Pakistan has a fairly wide-ranging national space programme covering many areas of space science, technology and applications. It is one of the medium-sized space programmes in the region, and is being implemented by the national space agency, the Pakistan Space and Upper Atmosphere Research Commission (SUPARCO). Pakistan was one of the earliest countries in the region to formally initiate a research programme concerning space and the upper atmosphere. SUPARCO was set up, initially as a Committee, as far as back as 1961, and the first sounding rockets for studying the upper atmosphere were fired in 1962. SUPARCO was elevated to the status of a fully-fledged autonomous commission under the federal government in 1980 through a Presidential Ordinance ratified by the national parliament. A high-powered Space Research Council (SRC), headed by the Prime Minister, functions as the apex body in matters concerning space. It lays down policy guidelines, approves the long-term space programme and gives the necessary directions to SUPARCO for its implementation. The SRC Executive Committee (ECSRC) approves the SUPARCO budget and monitors the progress of the space programme on behalf of SRC. With an established institutional framework and facilities relating to space, and the country's large human resource base, which includes a growing number of qualified and trained personnel, the national space programme is expected to expand further in the coming years.

In addition to this three-tier structure of SRC, ECSRC and SUPARCO, there are broad-based technical advisory committees in the country, comprising representatives from relevant departments or organizations, to assist SUPARCO in formulating its research programmes. Further, in 1994, the government set up the Pakistan National Coordination Committee, consisting of several concerned departments and organizations, including SUPARCO, to coordinate activities related to RESAP. Another national committee has been established for the effective implementation of the satellite-aided search and rescue COSPAS-SARSAT programme.

Pakistan's space programme encompasses diverse scientific and technological fields, which include research in atmospheric and space sciences using satellites and sounding rockets, satellite meteorology, remote sensing for resource and environmental surveying, atmospheric pollution, use of satellites to receive data from remote unattended DCPs for environmental studies, ionospheric physics and radio wave propagation, development of small satellites for communications, earth observations and other applications, establishment and operation of ground stations for remote sensing, meteorological and communication satellites, and astronomy. SUPARCO gives due priority to remote sensing. A national remote sensing applications centre, RESACENT, has been functioning at Karachi since 1978, while a ground receiving station for Landsat, SPOT and NOAA HRPT data has been operational at Islamabad since 1988-1989. The station was modernized and upgraded in 1997, which resulted in a significantly improved throughput capacity, generation of data products in new media and storage of archived data in more compact media.

Pakistan's first small experimental satellite, BADR-1, carrying a communication experiment, was indigenously designed and fabricated; it was launched from a Chinese Long March-2E vehicle in 1990. The next satellite, BADR-B, is now in an advanced stage of development and is expected to be launched towards late 1997. It has been decided that the planned national operational satellite communications system, PAKSAT, will be mainly handled by the private sector. SUPARCO has also set up a local user terminal and Mission Control Centre for the COSPAS-SARSAT programme at Lahore.

Besides SUPARCO, various other organizations in the country are involved in different space-related activities. The Pakistan Telecommunication Corporation, for instance, is responsible for regulating the development and operation of the telecommunications sector, in which space communications play a major role. Two Earth stations linked to Intelsat are operating near Karachi and Islamabad. The Pakistan Meteorological Department uses meteorological satellite data and balloons, radiosondes and precipitation radar for weather forecasting. In the field of remote sensing applications, more than 60 government, semi-government and private sector user agencies are applying satellite data disseminated by the SUPARCO ground station in different resource and environmental surveying and management projects. Regular courses in space science and space physics at Bachelor and Master degree levels are conducted, with help from SUPARCO scientists, at two of the leading universities in the country. SUPARCO also conducts short training courses in remote sensing regularly for other user agencies.

Further information on Pakistan's space programme may be obtained from:

Mr Abdul Majid
Chairman
SUPARCO
P.O. Box 8402
Karachi 75270
Pakistan
Fax: (92 21) 496-0553
E-mail: suparco@biruni.erum.com.pk

General Information on National Space Activities

National space agency responsible for coordination of space activities	Pakistan Space and Upper Atmosphere Research Commission (SUPARCO).
Mandate	SUPARCO was set up as a Committee in September 1961 with the objective of acquiring the capability to make use of space science and technology for peaceful purposes, primarily aimed at the socio-economic uplift of the people of Pakistan. SUPARCO was completely re-organized as an autonomous national organization with the status of a fully-fledged commission under the Federal Government, through Presidential Ordinance No. XX of 21 May 1981. Later ratified by the Parliament. Under this Ordinance, a Space Research Council (SRC) and an Executive Committee of the Space Research Council (ECSRC) were also constituted.
	SRC, headed by the Prime Minister, is the supreme body which directs and controls the space science and technology programmes in Pakistan, and approves its long-term programmes. ECSRC oversees the affairs of SUPARCO on behalf of SRC.
	As the national space agency, SUPARCO is responsible for the implementation of the national space programme approved by SRC. Pakistan's national space programme is essentially aimed at furthering research in space science and related fields, enhancing capabilities in space technology and promoting the applications of space science and technology for the socio-economic development of the country.
Organizational set-up	See figure 3.14.
Brief outline of the national space programme	Pakistan's space programme covers many fields of space science, technology and applications. It was initiated in 1962, when SUPARCO joined the Indian Ocean Expedition by firing its first sounding rocket, Rehbar-l. The national space programme now encompasses, among others:

(a) Development of facilities for the design, fabrication and launching of sounding rockets for upper atmosphere research, light-weight satellites in near Earth orbit;

(b) Establishment and operation of ground receiving stations for the acquisition of data for Earth resources and environmental surveying as well as for the collection of data from remote unmanned platforms under the Argos programme or receiving distress signals under the COSPAS-SARSAT programme;

(c) Establishment and operation of facilities for the tracking of satellites/rockets to determine their orbital parameters;

P

(d) Studies in space and atmospheric sciences, including tropospheric-stratospheric studies, environmental pollution monitoring, satellite geodesy, and astronomy;

(e) Studies based on application of satellite remote sensing data and geographic information system (GIS) to natural resources surveying and mapping and environmental monitoring;

(f) Studies relating to the ionosphere and associated radio wave propagation and geomagnetism.

Organizations responsible for various facets of the national space programme

SUPARCO is the sole national space agency. It has primary responsibilty for the implementation and coordination of various facets of the national space programme.

Legislation, government policy or directive related to integration of space applications in the development planning process

The Pakistan National Coordination Committee for RESAP was constituted, following a directive of the Federal Government, with the Cabinet Secretary, Cabinet Division, Government of Pakistan, as its chairman, and SUPARCO as the coordinating agency. It draws its membership from a number of relevant organizations besides SUPARCO, such as the Pakistan Meteorological Department, Pakistan Agriculture Research Council, Oil and Gas Development Corporation, Survey of Pakistan, Pakistan Environment Protection Agency, Water and Power Development Authority and World Conservation Union.

Coordination and promotional activities

SUPARCO is the national coordinator for remote sensing activities in Pakistan, in which capacity it is playing a pivotal role in promoting the use of remote sensing in the country. The SUPARCO Remote Sensing Applications Centre, established at Karachi in 1978, functions as the country's national remote sensing applications centre. There are also two technical advisory committees set up by the Government of Pakistan, one for satellite communications and space research and the other for the Earth Observations Programme to assist SUPARCO in formulating its research programmes in these fields. These technical advisory committees draw their membership from senior repreentatives of a number of national organizations.

Status of space science and technology education

To take care of its increasing manpower needs in various specialized fields of space science and technology, SUPARCO is in the process of setting up an aerospace institute in the premises of its satellite ground station at Islamabad. The postgraduate courses at the Aerospace Institute will be of two years' duration, leading to the award of a Master's degree in Space Technology and an M.Phil. degree in Space Science.

SUPARCO also provides various universities in the country with support in the conduct of courses relating to space science/space physics/remote sensing at the Bachelor's and Master's levels under the ongoing collaboration between SUPARCO and these universities.

SUPARCO provides teaching support for the Physics and Geography departments of the University of Karachi in courses relating to space physics/remote sensing at the Master's level. Similarly, SUPARCO scientists deliver lectures on space science to B.Sc. and M.Sc. students of the Department of Space Science of the University of the Punjab, Lahore. SUPARCO has, in fact, assisted the University in the establishment of this Department.

Regional space applications programme network

Contact	Name	Agency
RESAP national focal point	Mr Abdul Majid	Pakistan Space and Upper Atmosphere Research Commission
Regional Working Group on Remote Sensing, GIS and Satellite-based Positioning	Mr Jawed Ali	Pakistan Space and Upper Atmosphere Research Commission
Regional Working Group on Satellite Communication Applications	Mr S.A.H. Abidi	Pakistan Space and Upper Atmosphere Research Commission

Contact	Name	Agency
Regional Working Group on Meteorological Satellite Applications and Natural Hazards Monitoring	Mr Z.R. Siddiqui	Pakistan Space and Upper Atmosphere Research Commission
Regional Working Group on Space Sciences and Technology Applications	Mr M. Ishaq Mirza	Pakistan Space and Upper Atmosphere Research Commission

Respondent to questionnaires

Field	Agency	Name
Remote sensing, GIS and satellite-based positioning	Pakistan Space and Upper Atmosphere Research Commission	Mr M. Ishaq Mirza
Satellite communication applications	Pakistan Space and Upper Atmosphere Research Commission	Mr S.A.H. Abidi
Meteorological satellite applications and natural hazards monitoring	Pakistan Space and Upper Atmosphere Research Commission	Mr M. Ishaq Mirza
Space sciences and technology applications	Pakistan Space and Upper Atmosphere Research Commission	M. Nasim Shah

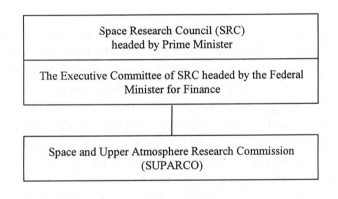

Figure 3.14 Organization of space activities in Pakistan

3.2.16 Philippines

The Philippines is engaged in a fairly wide-ranging space applications programme, which had its beginnings in the 1970s. Over the years, the national space effort has continued to expand in terms of size and scope, and encompasses four broad application areas, namely, remote sensing and GIS, satellite-based positioning, satellite meteorology and satellite communications, involving a large number of government and private sector agencies. The space programme has the necessary ingredients, such as infrastructure and facilities, qualified and experienced manpower, resources, and institutional framework, to expand and diversify further.

It was recommended at the Ministerial Conference on Space Applications for Development, held at Beijing in 1994, that a national coordinating body be set up in each country to oversee space applications. The National Coordinating Committee for Remote Sensing was therefore converted into the Science and Technology Coordinating Committee – Committee on Space Technology Applications (STCC-COSTA) to cover the whole range of space applications, inclusive of remote sensing, dealt with by RESAP. All space-related activities are now coordinated by STCC-COSTA, which comprises 16 national organizations. These organizations, which include the National Mapping and Resource Information Authority (NAMRIA), Department of Environment and Natural Resources, Bureau of Soils and Water Management, Philippine Atmospheric, Geophysical and Astronomical Services Administration (PAGASA), University of the Philippines, and National Telecommunications Commission, are engaged in activities that are relevant to the national space programme. STCC-COSTA functions under the Philippine Council for Advanced Science and Technology Research and Development of the Department of Science and Technology, which chairs its meetings.

The government accords high priority to remote sensing. Besides NAMRIA, which plays a pivotal role in development of remote sensing, about 20 other government and private sector agencies are involved in the application of remote sensing to a wide variety of resource and environmental studies, which include, for instance, forest resources monitoring, coastal zone management, coral reef mapping, groundwater assessment, geological mapping, lineament analysis of earthquake-prone areas and natural hazards mapping in the wake of the Mt. Pinatubo eruption. Some projects were carried out jointly with agencies in Australia and Germany, which provided a boost to the development of remote sensing capabilities in the country. Recently, there has been growing use of GIS in support of remote sensing. An inter-agency task force to promote and coordinate GIS use and standardization was set up in 1993. New projects relating to mapping and resource information, based on remote sensing and GIS, are being planned.

GPS has been used by NAMRIA to establish a geodetic control network across the country. GPS is being increasingly used for surveying, mapping and navigation purposes.

Data from NOAA and GMS meteorological satellites are used by PAGASA for weather analysis and forecasting, tracking of tropical cyclones, precipitation and flood forecasting, and monitoring of vegetation index and sea surface temperatures.

The Philippines is among the countries that have, or will soon have, their own national satellite communications systems. Two satellites, Agila and Mabuhay, built by France's Aerospatiale and the United States Space Systems/LORAL, respectively, were planned for launch in 1997-1998 to meet the growing domestic demand for communication services. The satellites will cover large areas of the Asian and Pacific region; thus, satellite transponders will be leased to Philippine users as well as to users in other countries.

Education and training in remote sensing/GIS and related areas is given due importance, with the University of the Philippines Training Centre for Applied Geodesy and Photogrammetry acting as the hub for such activities. It offers Master's degree and diploma courses in remote sensing and short training courses in remote sensing and GIS. In some other universities and colleges, remote sensing and GIS are part of courses in geography, environment and urban planning.

Further information on the Philippine space programme may be obtained from the RESAP national focal point:

Ms Ida Dalmacio
Executive Director
Philippine Council for Advanced Science and Technology Research and Development (PCASTRD) and Chairman, STCC-COSTA
DOST Compound, General Santos Avenue
Bicutan, Tagig, Metero Manila ZC 1631
Philippines
Tel: (63 2) 837-3171 to 90
Fax: (63 2) 837-2071 to 82

General Information on National Space Activities

National space agency responsible for coordination of space activities	Science and Techonology Coordinating Committee-Committee on Space Technology Applications (STCC-COSTA).
Mandate	The mandate of STCC-COSTA is to promote coordination, information exchange and collaboration and to function as a central body that will oversee the activities in the field of space technology applications and related fields in the country.
	The erstwhile National Coordinating Committee for Remote Sensing was converted into STCC-COSTA by virtue of STCC Resolution No. 4 Series of 1995.
	A number of organizations are members of STCC-COSTA, namely the National Mapping and Resource Information Authority, Philippine Atmospheric, Geophysical and Astronomical Services Administration, University of the Philippines Training Centre for Applied Geodesy and Photogrammetry, Department of Environment and Natural Resources, Philippine Institute of Volcanology and Seismology, Advanced Science and Technology Institute, Department of National Defence, Department of Transport and Communication, Philippine Council for Aquatic and Marine Research and Development, National Telecommunication Commission, National Security Council, PETEF, Department of Agriculture/Bureau of Soils and Water Management, Philippine Institute of Environment Planners, and Chamber of Mines of the Philippines. The Department of Science and Technology, represented by the Philippine Council for Advanced Science and Technology Research and Development, chairs the Committee.
Organizational set-up	See figure 3.15.
Brief outline of the national space programme	Applications of remote sensing to mapping, land-use planning, resource management, disaster mitigation, etc.; satellite communications; and meteorological applications.
Organizations responsible for various facets of the national space programme	See COSTA directory.
Legislation, government policy or directive related to integration of space applications in the development planning process	None.
Coordination and promotional activities	
Status of space science and technology education	

P

Regional space applications programme network

Contact	Name	Agency
RESAP national focal point	Ms Estrella F. Alabastro	Philippine Council for Advanced Science and Technology Research and Development
Regional Working Group on Remote Sensing, GIS and Satellite-based Positioning	Professor Epifanio D. Lopez	UP Training Centre for Applied Geodesy and Photogrammetry
Regional Working Group on Satellite Communication Applications	Ms Aurora A. Rubio	Department of Transportation and Communications
Regional Working Group on Meteorological Satellite Applications and Natural Hazards Monitoring	Mr Cipriano Ferraris	Philippine Atmospheric, Geophysical and Astronomical Services Administration
Regional Working Group on Space Sciences and Technology Applications	Mr Victorio Ochave	Advanced Science and Technology Institute

Responder to questionnaires

Field	Agency	Name
Remote sensing, GIS and satellite-based positioning	Cybersoft Integrated Geoinformatics, Inc.	Mr Oliver Coroza
	Geodata Systems Technologies, Inc.	Mr Elson Q. Aca
	Marine Science Institute	Mr Reuben Campos
	National Mapping and Resource Information Authority	Ms Virginia Sicat-Alegre
	National Water Resources Board	Ms Alma V. Nepomuceno
	UP Training Centre for Applied Geodesy and Photogrammetry	Mr Randy John N. Vinluan
Satellite communication applications	–	–
Meteorological satellite applications and natural hazards monitoring	Philippine Atmospheric, Geophysical and Astronomical Services Administration	Ms Flaviana Hilario and Mr Carmelito P. Calimbas
Space sciences and technology applications	–	–

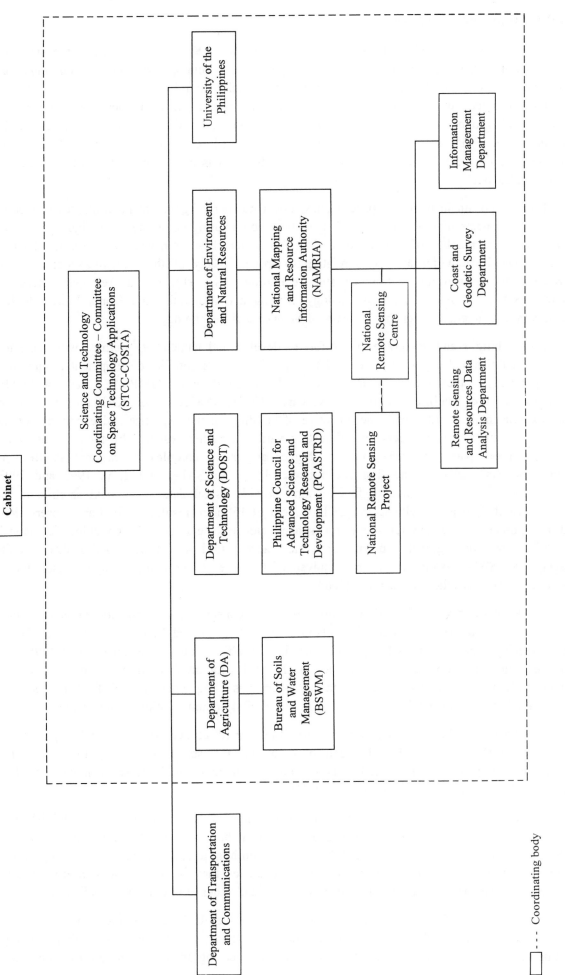

Figure 3.15 Organization of space activities in the Philippines

- - - Coordinating body

P

3.2.17 Republic of Korea

The economic and industrial strength of the Republic of Korea is reflected in the evolution of the national space programme. From fairly modest beginnings, the space programme has continued to expand in terms of size, scope and diversity, and is now on the threshold of a new era characterized by a significantly enhanced space segment with corresponding developments in the ground segment. The long-term programme of the Republic of Korea in the domain of space has set ambitious targets. It is planned to design, develop and launch 19 satellites over the next 20 years. A sum of US$ 6 billion has been approved by the government for this endeavour.

There is no single national space agency, but a number of government ministries concerned with various aspects of space are part of a national coordinating committee (see figures 3.16 and 3.17). A large number of organizations function under these ministries and are involved in different specific activities and tasks relating to space. Many of these organizations, such as Korea Advanced Institute of Science and Technology (KAIST), Korea Aerospace Research Institute (KARI), Korea Institute of Geology, Mining and Materials, and Korea Meteorological Administration, function under the Ministry of Science and Technology.

This ambitious and comprehensive effort to develop and launch 19 satellites will make use of the capabilities and expertise developed through three major satellite development programmes being pursued in the country, namely, KITSAT, Koreasat and KOMPSAT. The KITSATs, developed by the Satellite Technology Research Centre (SaTReC) of KAIST, are basically small experimental satellites with the objective of building up the expertise and experience of the country's scientists and engineers and of testing various systems designed by them in the actual environment of space, in preparation for bigger and more complex operational satellites in future. SaTReC launched the first two microsatellites, KITSAT-1 and 2, in 1992 and 1993, respectively, and these continue to function and provide useful data. The next satellite in the series, KITSAT-3, with a multispectral CCD camera of 15-metre resolution for Earth observations and a high-energy particle telescope for space science research, was scheduled for launch in 1997 to provide regional countries with easy access to low-cost but good-quality satellite imagery. KITSAT-4 would carry a more advanced CCD sensor providing a higher resolution of 5 metres as well as other scientific payloads.

The Koreasat system, funded by Korea Telecom, consists of two operational geostationary communication satellites, developed in collaboration with Lockheed Martin of the United States, with launchings carried out by McDonnell Douglas. Both satellites were launched in 1995-1996 and are being used operationally. Although the main communication needs have been met commercially through Intelsat, the Koreasat system has complemented and further broadened the range of services available, such as VSATs and digital communication broadcasting.

KOMPSAT would be a multipurpose satellite, aimed at acquiring national capability in developing satellites for land and ocean observations and space science applications. The programme is implemented by KARI in collaboration with major national space industries. It is planned to be launched around the middle of 1999.

A ground receiving station with a 13-metre antenna has recently been established at SaTReC. It will initially receive SPOT and JERS data, but will have the capability to acquire ERS and Radarsat data as well. KARI is also building a satellite remote sensing ground receiving station. In addition, there are several agencies that are receiving data from NOAA and GMS for weather forecasting, disaster monitoring and other environmental applications.

There has been fairly extensive use of data from the dedicated remote sensing satellites like Landsat, SPOT, JERS and MOS in the country. Major application areas include geology, mining, groundwater exploration, forestry, land use, cartography, oceanography and water resources. The availability of real-time data should provide a significant fillip to these various activities, leading to fully operational use of satellite remote sensing technology in the country.

Further information on the Republic of Korea's space programme may be obtained from the RESAP national focal point:

Mr Soon Dal Choi
Director
Satellite Technology Research Centre (SaTReC)
Korea Advanced Institute of Science and Technology
400 Kusong-dong, Yusong-gu, Taejon 305-701
Republic of Korea
Fax: (82 42) 861-0523, 861-0064
E-mail: sdchoi@krsc.kaist.ac.kr

Regional space applications programme network

Contact	Name	Agency
RESAP national focal point	–	Ministry of Science
Regional Working Group on Remote Sensing, GIS and Satellite-based Positioning	Mr Young-Kyu Yang	Systems Engineering Research Institute
Regional Working Group on Satellite Communication Applications	–	–
Regional Working Group on Meteorological Satellite Applications and Natural Hazards Monitoring	–	–
Regional Working Group on Space Sciences and Technology Applications	Professor Soon Dal Choi	Satellite Technology Research Centre

Respondent to questionnaires

Field	Agency	Name
Remote sensing, GIS and satellite-based positioning	Systems Engineering Research Institute	Mr Tae-Yong Kwon
Satellite communication applications	–	–
Meteorological satellite applications and natural hazards monitoring	Korea Meteorological Research Institute	Ms Ae-Sook Suh
Space sciences and technology applications	Satellite Technology Research Centre	Mr Taejung Kim

R

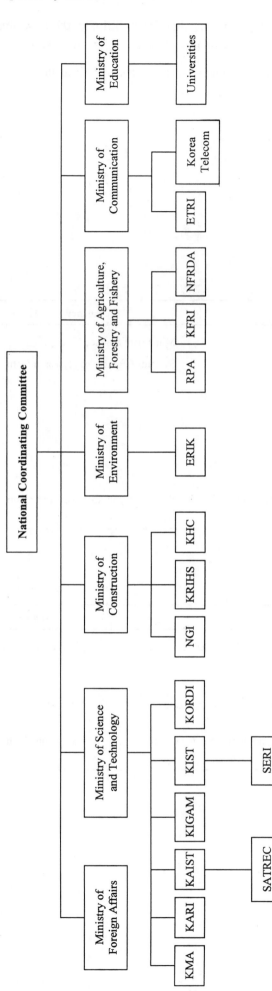

KMA　　 : Korea Meteorological Administration
KARI　　: Korea Aerospace Research Institute
KAIST　 : Korea Advanced Institute of Science and Technology
SATREC : Satellite Technology Research Centre
KIGAM　: Korea Institute of Geology, Mining and Materials
KIST　　: Korea Institute of Science and Technology
SERI　　 : System Engineering Research Institute
KORDI　 : Korea Ocean Research and Development Institute
NGI　　 : National Geography Institute
KRIHS　 : Korea Research Institute for Human Settlement
KHC　　: Korea Highway Corporation
ERIK　　: Environment Research Institute of Korea
RPA　　 : Rural Promotion Administration
KFRI　　: Korea Forestry Research Institute
NFRDA　: National Fishery Research and Development Agency
ETRI　　 : Electronic and Telecommunications Research Institute

Figure 3.16　Organization of space activities in the Republic of Korea

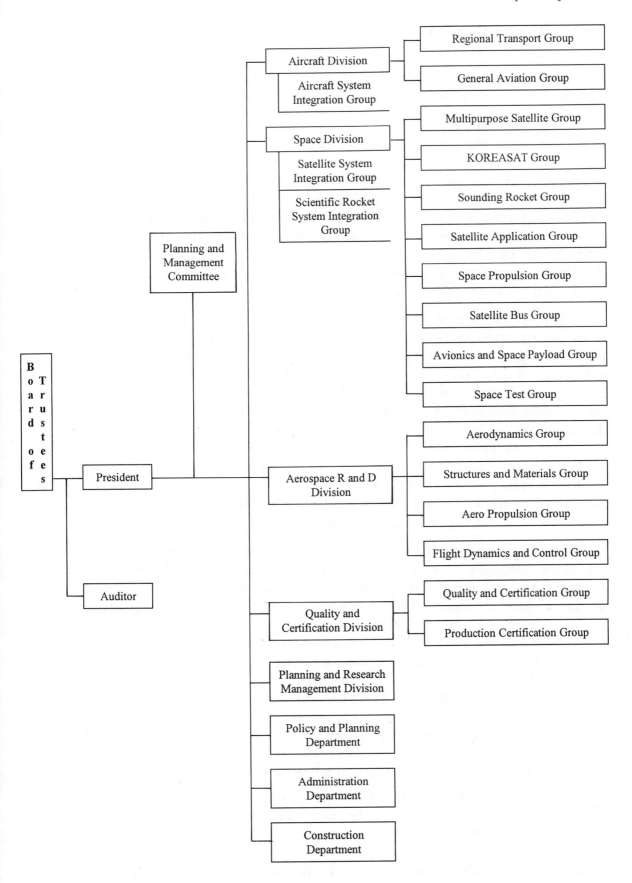

Figure 3.17 Organization chart of the Korea Aerospace Research Institute

3.2.18 Russian Federation

The Russian Federation was a part of the former Union of Soviet Socialist Republics, or Soviet Union, which sent the first artificial satellite, Sputnik-1, into orbit around the Earth in October 1957, thereby ushering in the "Space Age".

The Russian Federation has been the main inheritor of the space programme, inclusive of satellites/spacecraft and ground segments, of the former USSR after its dissolution in 1991. The various satellites launched during the Soviet era, such as, for instance, the Meteor series for meteorology, the Molniya, Raduga and Gorizont series for communications, the satellites for reconnaissance/mapping, and the GLONASS system for navigation, continue to be operated by the Russian Federation. Also, some of the space facilities being operated and used by the Russians are located in other countries that were once constituent republics of the former Soviet Union but are now independent sovereign countries, notably the Baikanour Cosmodrome in the Kazakhstan. Under an agreement with Kazakhstan, the Russian Federation will be able to use the facilities at Baikanour for a period of 20 years until 2014.

Currently, the Russian Federation (as well as many of the other constituent republics of the former Soviet Union that emerged or re-emerged as sovereign countries in 1991) are passing through a difficult economic period as its makes a transition from a highly controlled and planned socialist economy to a market economy. As a result of this current economic difficulty, the Russian Federation has not been able to allocate too many of its resources for further development of the space programme. Nevertheless, much of the expertise developed over the decades in the domain of space still exists in Russia. In fact, the Russian Federation has been offering various services, like high resolution remote sensing data from its satellites or the launching of satellites from its launchers, on a commercial basis to other countries. There has been a fairly positive response to these offers as many countries and organizations around the world are ordering Russian remote sensing satellite data or are signing contracts for the launch of their satellites through Russian launch vehicles.

The national space programme is being mainly implemented by the Russian Space Agency and various companies subordinate to or associated with it. Despite the severe economic and financial constraints due to the transition from a controlled to a free market economy, as mentioned above, the Russian Federation has managed to continue with the development and launch of some satellites like GOMS, their first geostationary meteorological satellite, the RESURS-1 for monitoring natural resources on land, and OKEAN-1 for ocean and ice monitoring in the Arctic and Antarctic regions. There are plans to launch more satellites in these series for land and ocean surveying, environmental monitoring and meteorology over the next three years. The principal satellite data acquisition and processing facilities are located in Moscow and its suburbs, in Novosibirsk and in Khabarovsk. In addition, there is a wide network of APT (automatic picture transmission) stations covering, besides the Russian Federation, former Soviet republics, as well as Antarctica. Since 1995, RESURS-1 data have been received at the Kiruna ground station in Sweden, which also markets the data.

As a whole, the Russian space effort is very substantial, covering a wide spectrum of areas in space science, technology and applications, and involving a large number of research organizations and contractors. The Soviet Union/Russian Federation is the only other country besides the United States to send manned missions into space. It is the only country to have maintained a regular orbiting space station – the Mir space station, which has provided invaluable data on many aspects of space travel and the space environment, including effects on humans. Designed to last five years when launched in 1986, Mir has continued to function quite satisfactorily for the last 11 years. Recently, a problem developed in the power system as a result of a docking manoeuvre, which has now been sorted out. Besides Russian crews, there have been several crews from the United States that have come and worked on Mir, laying a sound foundation for future cooperation between the two countries in space. Some individual astronauts from other countries, such as India, Mongolia and Viet Nam, have also been taken on Russian spacecraft. The Soviet Union/Russian Federation and the United States also share the distinction of being the only two countries so far to have sent (unmanned) spacecraft that successfully landed on the surface of the Moon and other planets, like Mars or Venus. However, despite the many "firsts" scored by the Soviet Union/Russian Federation in the space race, such as the first ever satellite or the first man to go into space, it has not yet landed men on the Moon (or any other celestial body), a feat that the United States accomplished through its Apollo missions. Another example of increasing international cooperation in space is the project to build a permanent international space station for supporting long-term experimentation and exploration in space. The five major partners in this effort will be the United States, Europe, the Russian Federation, Japan and Canada.

Further information on the Russian Federation's space programme may be obtained from the RESAP national focal point:

Mr A.I. Medvedchikov
Deputy General Director
Russian Space Agency (RSA)
42 Schepkina Street
Moscow 129090
Fax: (09 5) 288-9063, 975-1167

R

3.2.19 Singapore

Over the last few decades, Singapore has developed steadily and rapidly, registering remarkable economic growth rates. It has been able to achieve per capita GDP incomes and living standards that are on a par with those of many developed countries.

In view of its small size in terms of both area and population, Singapore does not need to pursue a large, ambitious and wide-ranging space programme, although it possesses much expertise in many modern science and engineering disciplines. Space technology applications in Singapore are thus mainly focused on satellite communications, satellite meteorology and remote sensing. Four major satellite earth stations linked to the global Intelsat satellite system, as well as the Palapa, AsiaSat, UoSat, KITSAT and PCSAT systems, are operating in Singapore and provide business, personal, financial, news and entertainment telecommunications services to the people. In addition, a coastal station has also been set up to receive Inmarsat data and provide a communications link for maritime vessels of all sizes as also for road vehicles and remote land- and sea-based sites.

Although various research groups in Singapore have been using remote sensing data since the 1980s, remote sensing received a boost in the country in 1992 with the establishment of the Centre for Remote Imaging, Sensing and Processing (CRISP) in the National University of Singapore (NUS) with funding from the National Science and Technology Board. NUS serves as the focal or lead agency for remote sensing activities in the country, representing Singapore in international and regional space applications programmes like RESAP. NUS offers courses in remote sensing and is engaged in research studies involving remote sensing in various disciplines like oceanography.

Soon after its establishment, CRISP set up a ground receiving station for satellite remote sensing data which became operational in September 1993. This station, which utilizes modern hardware and software, is one of the most sophisticated stations in the ESCAP region. It is acquiring data from SPOT and ERS, and is expected to start receiving Radarsat soon.

Scientists at Nanyang Technological University are now enganged in the fabrication and testing of small satellites, sounding rockets, telemetry equipment and launch facilities.

Satellite meteorology activities are carried out by the Meteorological Service Singapore, which is mandated to provide meteorological services for social and economic welfare, collaborate with regional and international meteorological agencies and monitor transboundary pollution. There is a meteorological satellite ground receiving station at Changi Airport, which receives and processes GMS and NOAA AVHRR data for routine weather forecasting. Notable projects the agency has undertaken include the development of a haze early warning system and retrieval of meteorological parameters from satellite data for integration with existing weather models. The country also provides host facilities to the ASEAN Specialized Meteorological Centre which carries out research studies relating to meteorology.

GIS is used in a large number of public and private sector organizations in Singapore. In 1989, the Singapore Land Data Hub was conceived. Its objective is to provide an integrated infrastructure for improved management of the the vast data resources that exist in the country to avoid unnecessary duplication and overlap. The hub model was adapted from a prototype study funded by UNDP. In its present form, the Land Data Hub plays the role of a central repository of geographic information with over 30 agencies contributing to it. Data ownership still rests on the contributing agency and data sharing is limited only to these contributing agencies. It is able to provide an efficient and cost-effective means of sharing and exchanging geographic information. It also reduces duplication and provides a consistent base for all other derivative data.

There is widespread growing interest in GPS in the country. At the government level, GPS may be used to establish a country-wide coordinate-based control system. It is expected that the use of GPS in such a system will result in a more consistent control framework for mapping and GIS activities.

Further information on Singapore's space programme may be obtained from the RESAP national focal point:

Mr Lim Hock
Centre for Remote Sensing Imaging, Sensing and Processing (CRISP)
National University of Singapore
Kent Ridge
Singapore 0511
Fax: (65) 792-3923
E-mail: crshefen@nus.edu.sg

General Information on National Space Activities

National space agency responsible for coordination of space activities	Singapore does not have a national space agency.
Mandate	
Organizational set-up	See figure 3.18.
Brief outline of the national space programme	There is no national space programme as such. However, some space-related activities, as listed below, are carried out such as reception and application in weather prediction of meteorological satellite data; reception, processing and application of SPOT, ERS and Radarsat data; and a joint project in satellite communication with the University of Surrey.
Organizations responsible for various facets of the national space programme	Meteorological Service Singapore; Centre for Remote Imaging, Sensing and Processing (CRISP), National University of Singapore; and School of Electrical and Electronic Engineering, Nanyang Technological University.
Legislation, government policy or directive related to integration of space applications in the development planning process	None.
Coordination and promotional activities	Some space-related activities are coordinated on an *ad hoc* basis by the National Science and Technology Board.
Status of space science and technology education	Various courses on remote sensing, GIS, satellite communications, etc., are conducted at undergraduate and postgraduate levels in tertiary educational institutions in Singapore.
	Many seminars organized either solely by CRISP or jointly with SPOT Asia on SAR, visible remote sensing study of forests/vegetation, ocean phenomena, urban studies, etc., as well as remote sensing data processing techniques. CRISP staff are sent on various local or international training programmes on remote sensing, computer management, GIS, data processing techniques, etc.

Regional space applications programme network

Contact	Name	Agency
RESAP national focal point	Mr Lim Hock	Centre for Remote Imaging, Sensing and Processing
Regional Working Group on Remote Sensing, GIS and Satellite-based Positioning	–	–

Contact	Name	Agency
Regional Working Group on Satellite Communication Applications	–	–
Regional Working Group on Meteorological Satellite Applications and Natural Hazards Monitoring	–	–
Regional Working Group on Space Sciences and Technology Applications	–	–

Respondent to questionnaires

Field	Agency	Name
Remote sensing, GIS and satellite-based positioning	Centre for Remote Imaging, Sensing and Processing	Mr Lim Hock
Satellite communication applications	Nanyang Technological University	–
Meteorological satellite applications and natural hazards monitoring	Meteorological Service Singapore	Mr Koo Hock Chong
Space sciences and technology applications	Nanyang Technological University	Mr Tan Soo Hie

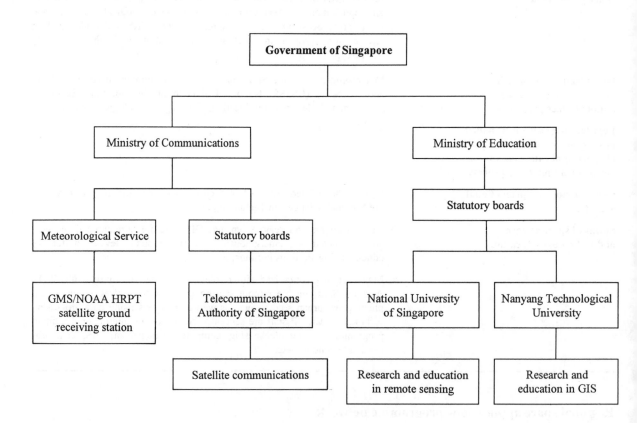

Figure 3.18 Organization of space activities in Singapore

3.2.20 Sri Lanka

Sri Lanka, one of the comparatively smaller countries of South Asia, is pursuing a modest-sized space programme that is essentially focused on remote sensing and associated GIS, satellite meteorology and satellite communications.

There is no national space agency, as such. The Arthur C. Clarke Centre for Modern Technologies, established in 1984 to accelerate the promotion, assimilation and development of modern technologies in Sri Lanka, has been designated the national coordinating agency for space applications in the country. The Centre is actively engaged in R and D studies, education and training activities, consultancy work and so forth, in fields related to computers, communications, electronics, space science and technology. The Centre is also in the process of establishing a section for satellite remote sensing and GIS. Further, it has taken the initiative to instal a 45-cm optical telescope, with Japanese assistance, for astronomical observations. These facilities will also be available to other government organizations as well as universities.

The National Survey Department is the principal agency in Sri Lanka involved in the application of satellite remote sensing data. Since the launch of the Landsats in the 1970s, the Survey Department has been making extensive use of the data for mapping. In fact, the Survey Department was one of the pioneers in the region as far as the use of Landsat data for mapping was concerned. It is presently using Landsat, SPOT and IRS data in its work. It has PCI and IDRIS image processing software packages on powerful micro-computers, which are being used for applications in land-use change, hydrological and forest cover mapping. The Survey Department has also conducted many training courses in remote sensing applications, which have greatly helped to promote the use of this technology in relevant oganizations in the country. There are now many government agencies, such as the Mahaweli Authority and the Forest Department, which are using satellite remote sensing data in their work. The Forest Department has established a satellite remote sensing section and is mapping forest cover at the 1:50,000 scale and overlaying it on the national topographical series at the same scale. Most of this mapping was done using IRS data. The Mahaweli Authority is executing a project to study land use, environment and forest conservation in the Mahaweli catchment areas using satellite data.

The Department of Meteorology has been using satellite data for weather forecasting since about 1972. It receives data from the polar-orbiting NOAA series of satellites and the geostationary Japanese GMS satellite over the western Pacific Ocean. India is expected to provide equipment, under the Indo-Sri Lanka Commission on Science and Technology, to receive INSAT data. Microcomputers-based signal processing systems are used for weather forecasting.

GIS are also presently being used by many agencies in the country. Notable work, in this regard, has been done by the University of Colombo in malaria control and the University of Peradeniya in coastal zone management. The Survey Department is working on converting topographic data into digital form for these applications. Two pilot GIS projects in planning have been carried out. The Mahaweli Authority has the largest GIS data base covering most of the Central Province.

Sri Lanka Telecom, the national organization responsible for satellite communications, operates three satellite earth stations. In addition, it has two sub-marine cable systems. About 60 per cent of the international telephone traffic is handled by the satellite stations. The stations are linked to different Intelsat satellites depending on the destination. Sri Lanka Telecom has four Inmarsat portable Earth stations for personal and voice communication applications. The Arthur C. Clarke Centre itself has also established a facility for utilizing an Amateur Satellite System to receive and transmit information from the Global Amateur Satellite Network with assistance from JICA. The country hopes to expand satellite communication specially for data transfer, education and training in the coming years.

S

The Government of Sri Lanka has shown its support for space technology. Following the Ministerial Conference held in Beijing in 1994, the government has ratified the Beijing Declaration, approved the establishment of a space applications centre within the Arthur C. Clarke Centre and also approved the organization of a national committee for space technology applications.

Further information on Sri Lanka's space programme may be obtained from the RESAP national focal point:

> Mr Sam Karunaratne
> Director
> Arthur C. Clarke Centre for Modern Technologies
> Katubedda, Moratuwa
> Sri Lanka
> Fax: (94 1) 507462
> E-mail: accmt@mail.ac.lk

General Information on National Space Activities

National space agency responsible for coordination of space activities	Arthur C. Clarke Centre for Modern Technologies.
Mandate	The Arthur C. Clarke Centre for Modern Technologies has the mandate to introduce and accelerate the development of activities relating to space sciences and technology, communications, computers, robotics and energy in Sri Lanka.
Organizational set-up	See figure 3.19.
Brief outline of the national space programme	The national space programme covers space science and technology, space communications, satellite remote sensing and GIS and satellite meteorology.
	National policies are being outlined by the National Committee headed by the Minister for Science Technology and Human Resources Development. The focal point for the space programme, the Arthur C. Clarke Centre, is coordinating the implementation of space activities in Sri Lanka.
Organizations responsible for various facets of the national space programme	Space sciences and satellite communications activites are carried out by the Arthur C. Clarke Centre for Modern Technologies. Remote sensing and GIS activities are carried out by the Department of Survey. Satellite meteorology and disaster monitoring activities are carried out by the Department of Meteorology.
Legislation, government policy or directive related to integration of space applications in the development planning process	A policy will be drafted very soon to integrate the space programme into the national development planning process.
Coordination and promotional activities	The Arthur C. Clarke Centre has coordinated activities relating to the training of personnel in different organizations during the past two years. It also assisted in organizing international conferences which were held in Sri Lanka and were sponsored by local and international organizations. It coordinates the activities of the National Committee on Space Applications.
Status of space science and technology education	Space sciences and technology education in the universities in Sri Lanka needs to be substantially improved. Few universities have commenced work in remote sensing and GIS. Some universities are training their personnel abroad in remote sensing and space communications courses.
Other pertinent information or remarks	The progress of space science and technology education has been particularly slow because of a shortage of manpower and resources required for building infrastructure.

Regional space applications programme network

Contact	Name	Agency
RESAP national focal point	Mr Sam Karunaratne	Arthur C. Clarke Centre for Modern Technologies
Regional Working Group on Remote Sensing, GIS and Satellite-based Positioning	Surveyor General	Survey Department
Regional Working Group on Satellite Communication Applications	Mr Sam Karunaratne	Arthur C. Clarke Centre for Modern Technologies
Regional Working Group on Meteorological Satellite Applications and Natural Hazards Monitoring	Director	Department of Meteorology
Regional Working Group on Space Sciences and Technology Applications	Mr Sam Karunaratne	Arthur C. Clarke Centre for Modern Technologies

Respondent to questionnaires

Field	Agency	Name
Remote sensing, GIS and satellite-based positioning	Environment and Forest Conservation Division, Mahaweli Authority	Mr H. Manthrithilake
Satellite communication applications	Arthur C. Clarke Centre for Modern Technologies	Mr H.S. Padmasiri de Alwis
Meteorological satellite applications and natural hazards monitoring	Department of Meteorology	Mr N.A. Amaradasa
Space sciences and technology applications	–	–

S

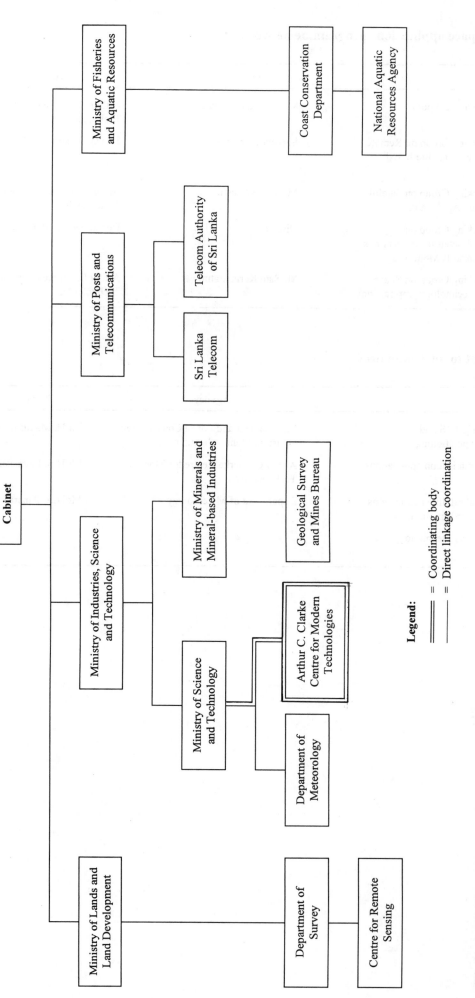

Figure 3.19 Organization of space activities in Sri Lanka

3.2.21 Thailand

Thailand is pursuing a fairly broad-based medium-sized space programme, which is larger in size and scope than that of most developing countries in the Asian and Pacific region. Thailand's space effort was initially focused on satellite remote sensing and its applications, which still remains an important component. With the passage of time, the space programme has become larger and more diversified and covers satellite meteorology, satellite communications, distance education and development of small satellites.

Thailand's active involvement in research relating to the use of space technology dates back to 1971, when the National Research Council of Thailand (NRCT) initiated the Thailand Satellite Remote Sensing Programme. The remote sensing programme in the country is well established and quite comprehensive. NRCT, as well as about 40 other user agencies in the country, are applying satellite remote sensing, often in conjunction with GIS, to various natural resources and environmental problems confronting the country, especially those relating to deforestation, coastal zones, floods, water resources management and land-use changes. In fact, the high economic growth rates witnessed in Thailand in recent years have put increasing pressure on natural resources and on the land, and has, in some ways, exacerbated environment problems. Hence, the use of remote sensing and GIS is even more relevant now for the country. The application of remote sensing data in the country has been greatly facilitated by their real time/near real time availability from the ground receiving station at Lat Krabang near Bangkok. This station established in 1981 to receive Landsat MSS data was one of the earliest ground stations in the Asia-Pacific region. Thailand thus possesses much experience and expertise in satellite remote sensing data acquisition, processing, archiving and dissemination. The station has been successively upgraded to accommodate newer satellites and now receives data from as many as seven satellite systems, namely, Landsat, SPOT, MOS, NOAA, ERS, JERS and Radarsat, the latest addition. It, perhaps, receives data from more satellite systems than any other ground station in the region. Data from meteorological satellites are also being routinely used for weather forecasting and natural disaster monitoring in Thailand.

A Space Development Committee has been formed under the Cabinet, while a Space Development Agency has also been established under the Ministry of Transport and Communications (MOTC), reflecting the increasing importance of space related activities in the country. NRCT, under the Ministry of Science, Technology and Environment, continues to be responsible for the remote sensing programme, while MOTC deals with other fields/aspects (see figure 3.20).

There has been, in particular, a significant expansion in satellite communications in the last three to four years to cater to rapidly growing demands in the government and business sectors, as also among individual consumers/users, fuelled largely by the high economic growth rates witnessed in Thailand for several years. A national satellite communications satellite system, called Thaicom, has been operational since 1993. Thaicom is managed by a private sector operator, Shinawatra Satellite Public Company, in line with the increasing involvement of private operators/service providers in telecommunications in the region. The first two satellites of the Hughes-built Thaicom series, Thaicom-1 and Thaicom-2, were launched by Arianespace in 1993 and 1994, respectively. The third, Thaicom-3, was launched in April 1997. These satellites are being used to provide a wide range of data communications, telephone, VSAT operations, broadcasting and other services by many government and private agencies.

Thailand has embarked upon a programme for the development of small satellites that could be used for various purposes, such as Earth observations, environmental monitoring, disaster monitoring and communication experiments. As a part of this programme, Thailand is planning the development of a dedicated remote sensing satellite, the Thai Remote Sensing Small Satellite, TRSSS, in cooperation with Canada. Also, a microsatellite called the Thai Microsatellite

T

is being jointly developed and built by a private Thai university, a private Thai company and a university in the United Kingdom.

Further information on Thailand's space programme may be obtained from the RESAP national focal point:

Mr Jirapandh Arthachinta
Secretary-General
National Research Council of Thailand
196 Phaholyothin Road
Bangkok 10900
Thailand
Fax: (66 2) 561-3035
E-mail: darasri@fc.nrct.go.th

Regional space applications programme network

Contact	Name	Agency
RESAP national focal point	Mr Jirapandh Arthachinta	National Research Council of Thailand
Regional Working Group on Remote Sensing, GIS and Satellite-based Positioning	Ms Darasri Dowreang	National Research Council of Thailand
	Capt Kanok Weerawong	Royal Thai Survey Department
Regional Working Group on Satellite Communication Applications	Mr Kosol Choochuay	Centre for Education Technology, Ministry of Education
	Ms Rasamee Suwanweerakamtorn	Thailand Remote Sensing Centre
Regional Working Group on Meteorological Satellite Applications and Natural Hazards Monitoring	Ms Dusadee Sarigabutr	Meteorological Department
Regional Working Group on Space Sciences and Technology Applications	Mr Suthi Aksornkitti	King Mongkut Institute of Technology

Respondent to questionnaires

Field	Agency	Name
Remote sensing, GIS and satellite-based positioning	Thailand Remote Sensing Centre	Ms Rasamee Suwanweerakamtorn
Satellite communication applications	Thaicom Distance Education Centre	Mr Sanong Chinnanon
	Post and Telegraph Department	–
	Telephone Organization of Thailand	–
	Communication Authority of Thailand	Mr Traichuk Prommakoun
Meteorological satellite applications and natural hazards monitoring	–	–
Space sciences and technology applications	–	–

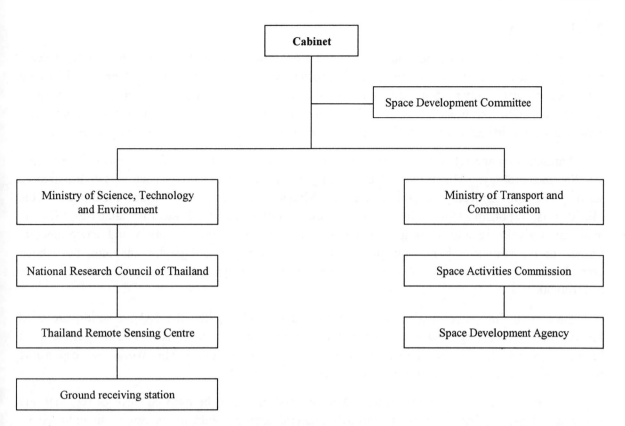

Figure 3.20 Organization of space activities in Thailand

3.2.22 Vanuatu

Vanuatu is a small island country situated in the south-western part of the Pacific Ocean. Although relatively larger, in terms of both population and area, than most other Pacific island countries, Vanuatu suffers from the drawbacks and handicaps that are typical of the countries of this subregion, such as geographic isolation from the continental land masses, small land area, inadequate infrastructure, small size of the economy and vulnerability to natural disasters.

Vanuatu's space-related activities are principally focused on remote sensing, GIS and satellite meteorology. The lead or focal agency in the country with respect to satellite remote sensing and GIS projects and activities is the Ministry of Lands and Natural Resources. This Ministry is responsible for maintaining a forest resource and land records database. Use is made of available remote sensing data and GIS in various natural resources and environmental studies in the country. Forest mapping and inventorying and topographic mapping have been carried out with support from Australia. Forest resource maps have been produced at a scale of 1:100,000.

Data from the meteorological and environmental satellites, like the NOAA series, are used for weather forecasting and for environmental management activities. The Department of Meteorology, which functions under the Ministry of Transport and Public Works, is responsible for weather forecasting and monitoring activities.

Like the other Pacific island nations, Vanuatu also needs to be assisted through regional and international programmes as well as through bilateral arrangements in its efforts to enhance its capacity to make use of space technology for various nation-building activities.

Further information on Vanuatu's space programme may be obtained from the RESAP national focal point:

Mr Edwin Arthur
Permanent Secretary
Survey Department
Ministry of Lands, Geology, Mines, Energy and Rural Water Supply
Private Mail Bag 007, Port Vila
Vanuatu
Fax: (67 8) 25051, 25165

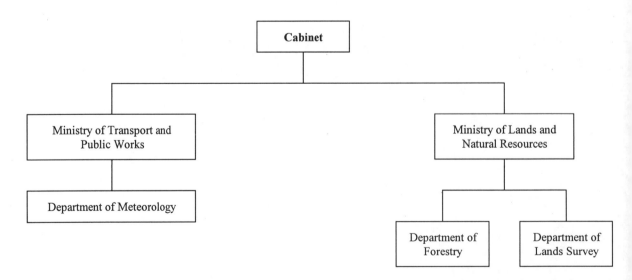

Figure 3.21 Organization of space activities in Vanuatu

3.2.23 Viet Nam

Viet Nam is one of the larger countries of South-East Asia in terms of population. As a developing country, Viet Nam is eager to achieve high economic growth and is engaged in a sizeable economic development and reconstruction programme, in which space technology is perceived as having a significant role.

Responsibility for coordination of space-related activities in Viet Nam rests with the National Council for Space Technology Application. This Council has been assigned the role of defining the direction and thrust of the national space applications programme, especially in the context of economic development. At present, over 20 institutions and organizations, including the Institute of Physics, Institute of Geography, Institute of Geology, Institute of Mining and Geology, Institute for Agriculture Planning, Institute for Forest Inventory and Planning, General Department of Geodesy and Cartography, Institute of Oceanography, General Department of Meteorology and Hydrology, and the Telecommunication Centre, and universities are engaged in space applications.

The space programme in Viet Nam is essentially concerned with the applications of remote sensing, GIS, satellite meteorology and satellite communications, as is the case with most developing countries. Current remote sensing applications focus on resource management and inventorying, and environmental protection and management. Data from Landsat, SPOT, IRS, Soyuz and other satellites, as well as aerial photographs, are used in various projects for integrated studies, including land-cover and land-use mapping, vegetation monitoring, geological and geotectonic mapping, environmental monitoring, coastal dynamics studies, soil surveying, ecosystem studies, groundwater investigations, sedimentation studies, flood mapping and oceanographic studies. Viet Nam has invested heavily to obtain aerial photography and Landsat and SPOT data of the whole country.

Increasing use is now being made of GIS in various organizations, including academic institutions, in the country in resource mapping, planning and management projects. A large number of organizations possess GIS facilities like Arc Info, ArcView, SPANS, ILWIS, PC 1 and others. Some GIS software has also been developed locally. GIS is being utilized to create digital data bases of existing topographic and thematic maps.

Realizing the need for training to keep up with the country's needs, several institutions are involved in providing degree programmes and training courses on remote sensing and GIS. These institutions include the Institutes of Physics, Geography and Geology, the Polytechnic Institute, Hanoi University, Can Tho University and the Institute of Hydrolautics and Water Resources.

Some meteorological satellite ground receiving stations have been set up in Viet Nam to receive and process images from meteorological satellites like GMS, NOAA and Meteor for routine weather forecasting and disaster monitoring, especially tropical cyclone warning. Some studies have been carried out using meteorological satellite data for chlorophyll concentration and sea surface temperature mapping. The data are also used for studying vegetation cover seasonal changes at small scale in the whole country. Another project to monitor disasters caused by the interaction between sea and land is currently being carried out over 1,000 km of coastline in central Viet Nam.

The country has established cooperation with Intelsat for telecommunications services in the country. In parallel with the installation of optic cable lines, a satellite communication network has been established for the entire country and has been connected to the international satellite network by the Thaicom satellite. Distance education is also being developed through radio and television broadcasts with several programmes on foreign languages, mathematics, literature, physics, chemistry and many other topics.

V

Further information on Viet Nam's space programme may be obtained from the RESAP national focal point:

Professor Tran Manh Tuan
Deputy General Director
National Centre for Science and Technology of Viet Nam
Nghia Do Tu Liem, Hanoi
Viet Nam
Fax: (84 4) 835-2483
E-mail: bdtrong@netnam.org.un

General Information on National Space Activities

National space agency responsible for coordination of space activities	National Centre for Science and Technology of Viet Nam.
Mandate	Implementation of the national space programme and coordination of space activities in the country.
Organizational set-up	See figure 3.22.
Brief outline of the national space programme	Application of remote sensing and GIS technology, satellite meteorology, satellite communications.
Organizations responsible for various facets of the national space programme	Remote sensing and GIS application and software development activities are carried out by the National Centre for Science and Technology of Viet Nam. The application of remote sensing in agriculture is done by the Institute of Agriculture Planning under the Ministry of Agriculture and Country Development. The application of remote sensing in forestry is done by the Forest Inventory and Planning under the Ministry of Agriculture and Country Development. The application of remote sensing in cartography is done by the General Department of Cadastry. Satellite meteorology activitites are carried out by the General Department of Meteorology and Hydrology of Viet Nam. Satellite communications activities are carried out by the General Department of Post and Telecommunication.
Legislation, government policy or directive related to integration of space applications in the development planning process	None.
Coordination and promotional activities	The National Centre for Science and Technology of Viet Nam provides the government with advice on space activities and coordinates between relevant agencies for the conduct of joint projects involving the use of remote sensing.
Status of space science and technology education	Remote sensing and GIS technology is now one of the subjects that can be studied at the Geographical Faculty of Hanoi University as well as at the Faculty of Geodesy and Cartography in Hanoi Institute of Mines and Geology.

Regional space applications programme network

Contact	Name	Agency
RESAP national focal point	Professor Tran Manh Tuan	National Centre for Science and Technology of Viet Nam
Regional Working Group on Remote Sensing, GIS and Satellite-based Positioning	Mr Bui Doan Trong	Institute of Physics
Regional Working Group on Satellite Communication Applications	Mr Nguyen Thanh Phuc	General Department for Post and Telecommunications

Contact	Name	Agency
Regional Working Group on Meteorological Satellite Applications and Natural Hazards Monitoring	Mr Hoang Minh Hien	Hydro-meteorological Service of Viet Nam
Regional Working Group on Space Sciences and Technology Applications	Mr Nguyen Quang Thinh	Institute of Physics

Respondent to questionnaires

Field	Agency	Name
Remote sensing, GIS and satellite-based positioning	National Centre for Science and Technology of Viet Nam	Mr Nguyen Van Hieu
Satellite communication applications	Department General of Posts and Telecommunications	Mr Nguyen Thanah Phuc
Meteorological satellite applications and natural hazards monitoring	National Centre for Hydro-meteorological Forecasting	Mr Hoang Minh Hien
Space sciences and technology applications	–	–

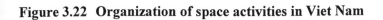

Figure 3.22 Organization of space activities in Viet Nam

Chapter 4

REMOTE SENSING, GIS AND SATELLITE-BASED POSITIONING

4.1 Regional overview

4.1.1 Background

The first dedicated remote sensing/Earth observation satellite, the United States' ERTS-1 (later renamed Landsat-1), was launched in 1972, about 12 years after the launch of the first operational meteorological satellites of the NOAA/TIROS series. However, it was remote sensing of Earth resources that, in a short span of time, became the best known and dominant element in the national space programmes of most countries. From the mid-1970s, remote sensing became the nucleus around which the space programmes of most developing countries in the region have developed. Remote sensing continues to be accorded high priority in national space programmes of developing countries because of its ability to address a variety of resource and environmental problems confronting these countries. There are about 30 national remote sensing centres in the region; more than 20,000 remote sensing specialists and scientists are engaged in over 2,000 projects based on the application of remote sensing data. These projects relate to diverse disciplines, such as agriculture, forestry, geology, soil science, mining, fisheries, hydrology, geography, oceanography, cartography, environment and ecology. In fact, in many developing countries, remote sensing often represents the main, if not the only, activity in the national space effort.

In ESCAP space-related collaborative programmes, remote sensing enjoys an important position relative to other space applications because of its relevance to the needs of developing countries. ESCAP executed, between 1983 and 1991, its Regional Remote Sensing Programme, which was exclusively devoted to remote sensing. Its follow-on project on integrated applications of geographic information systems and remote sensing for sustainable natural resources and environment management (GIS-RSRP) focused on the complementary use of remote sensing and GIS. In the current Regional Space Applications Programme for Sustainable Development (RESAP), remote sensing also figures prominently in the programme's operational activities as it can contribute directly to sustainable development. Of the four regional working groups formed under RESAP, the one concerned with remote sensing and GIS (and satellite-based positioning) has been given a pre-eminent position, as its annual meetings are convened back-to-back at the same venue as the annual sessions of the RESAP Intergovernmental Consultative Committee (ICC) and the annual meetings of the inter-agency Subcommittee on Space Applications for Sustainable Development. The national contact points for the Regional Working Group on Remote Sensing, GIS and Satellite-based Positioning are listed in table 4.1.

4.1.2 The remote sensing space segment

Remote sensing satellites, which typically orbit at altitudes between 700 and 900 km, have provided, over the last quarter of a century since the launch of Landsat-1, vast amounts of synoptic and repetitive data of the Earth's surface that are used in numerous applications. Over the years, remote sensing technology has continued to advance in several ways, including improved spatial and radiometric resolutions, greater use of spectral regions other than the optical, and oblique viewing capabilities. Spatial resolutions of 10, 8 and 6 metres provided by the French SPOTs, Japanese ADEOS and Indian IRS-1C respectively represent a significant advance over the 80-metre resolution of Landsat MSS and even the 30 metres of TM on later Landsats. Resolutions are, in general, expected to keep improving. For instance, SPOT-5, due for launch around 1999, will have 5-metre resolution, while Japan's advanced satellite dedicated

Table 4.1 National contact points of the Regional Working Group on Remote Sensing, Geographic Information Systems and Satellite-based Positioning

Country	Name	Agency
Australia	Mr David L.B. Jupp	CSIRO Office of Space Science and Applications
Azerbaijan	–	–
Bangladesh	Mr Anwar Ali	Space Research and Remote Sensing Organization
China	Ms Zheng Lizhong	National Remote Sensing Centre of China
Fiji	Mr M. Jaffar	Ministry of Lands, Mineral Resources and Energy
India	Mr V. Jayaraman	Earth Observation Systems Programme Office
Indonesia	Mr Bambang Tejasukmana	National Institute of Aeronautics and Space
Islamic Republic of Iran	Mr Ahmad Mohammadpour	Iranian Remote Sensing Centre, Ministry of Post, Telegraph and Telephone
Japan	Mr Akira Noie	Office of Research and International Affairs on Space Activities, Science and Technology Agency
Malaysia	Mr Nik Nasruddin Mahmood	Malaysian Centre for Remote Sensing, Ministry of Science, Technology and the Environment
Mongolia	Mr Sodov Khudulmur	Mongolian National Remote Sensing Centre
Myanmar	–	–
Nepal	Mr Rajendra B. Joshi	Forest Research and Survey Centre
Pakistan	Mr Jawed Ali	Space and Upper Atmosphere Research Commission
Philippines	Mr Virgilio S. Santos	National Mapping and Resource Information Authority
Republic of Korea	Mr Young-Kyu Yang	Systems Engineering Research Institute
Russian Federation	–	–
Singapore	–	–
Sri Lanka	Surveyor General	Survey Department
Thailand	Ms Darasri Dowreang	National Research Council
Vanuatu	–	–
Viet Nam	Mr Bui Doan Trong	Institute of Physics, National Centre for Science and Technology

to land applications, ALOS, expected to be launched in 2002, will provide 2.5 metre resolution. The Russian Federation is offering, on a commercial basis, very high-resolution satellite data of the order of 2 to 6 metres. Some private companies and consortia in the United States have developed remote sensing satellites for civilian use with very high resolutions of the order of 1 to 3 metres, expected to be launched during the period 1997-1999. This would greatly enhance the utility of civilian remote sensing satellites for detailed studies and substantially improve mapping accuracy. Other regions of the electromagnetic spectrum besides the optical, such as the middle infra-red, thermal infra-red and microwave, are now also being increasingly exploited. The 7-band TM has middle and thermal infra-red bands, while a number of dedicated satellites with microwave/radar imaging devices, such as the European Space Agency's ERS-1 and 2, Japan's JERS-1, United States/French Topex/Poseidon and Canada's Radarsat, have been launched. Since microwaves can penetrate clouds, these satellites are particularly useful for the tropical areas of the Asian and Pacific region that are often cloud-covered and, hence, cannot be

Table 4.2 Existing facilities and infrastructure in the Asian and Pacific region

Agency/country	Computer			Peri-pherals	GPS	Network	Image processing	GIS	Database
	MF	WS	PC						
CSIRO Office of Space Sciences and Applications, Australia	✓	✓	✓		Receivers and base stations	LAN, WAN, Internet	ERDAS	Arc Info, Idrisi, ILWIS, Intergraph, Mapinfo	Oracle
Azerbaijan National Space Agengy, Azerbaijan			✓	Scanner			(Custom)	(Custom)	Custom
National Remote Sensing Centre-Changsha, China			✓	Scanner, digitizers, printers			ERDAS, PCI	Arc Info, MapCAD, MapGIS, Mapinfo	dBase, Oracle
Wuhan Technical University of Surveying and Mapping, China		✓	✓	Scanner, digitizers, printers	Receivers and base stations	LAN, Internet	ERDAS	Arc Info, Intergraph, Mapinfo, SPANS	dBase, Foxpro, Oracle
Ministry of Lands, Mining and Energy, Fiji	✓	✓	✓	Digitizers, printers	Receivers	LAN, WAN		Intergraph, Mapinfo	Oracle, AREV
University of South Pacific, Fiji			✓	Digitizers, printers	Receivers	LAN, Internet		Arc Info, Idrisi, Osumap	Oracle, Access
National Institute of Aeronautics and Space, Indonesia	✓	✓	✓	Digitizers, printers, film writer	Receivers	Internet	ERDAS, PCI, ER Mapper, Meridian	Arc Info, Idrisi	Oracle
Iranian Remote Sensing Centre, Islamic Republic of Iran			✓	Digitizers, printers, film writer	Receivers	LAN, Internet	ER Mapper, PCI	Arc Info, Idrisi, ILWIS, Mapinfo, SPANS	dBase, Foxpro
National Space Development Agency, Japan		✓	✓	Digitizers, printers		LAN	ERDAS, PCI, ER Mapper	Arc Info, Vista/ Prism, WIT/Map	Oracle, Sybase
Ministry of International Trade and Industry, Japan		✓	✓	Digitizers, printers, film writer	Receivers and base stations		Envi, ER Mapper, TNTmips	Arc Info	
National Remote Sensing Centre, Mongolia	✓	✓	✓	Scanner, digitizers, printers, film writer	Receivers	LAN, WAN, Internet	ERDAS, PCI	Arc Info, Idrisi, ILWIS	dBase, Foxpro
Department of Meteorology and Hydrology, Myanmar		✓		Thermal printer					
Forest Research and Survey Centre, Nepal			✓	Digitizer, plotter		LAN	ERDAS	Arc Info, Topos	Foxpro
Space and Upper Atmosphere Research Commission, Pakistan	✓	✓	✓	Scanner, digitizers, printers, film writer	Receivers		EBBA, ERDAS, PCI	Arc Info	dBase, Oracle, Foxpro
Cybersoft Integrated Geoinformatics, Inc., Philippines		✓	✓	Scanner, digitizers, printers		LAN, Internet	ER Mapper	Genamap, Mapinfo	Foxpro, Informix
Geodata Systems Technologies Inc., Philippines		✓	✓	Digitizers, printers		LAN	ERDAS	Arc Info and other ESRI products	Foxpro

Table 4.2 *(Continued)*

Agency/country	Computer			Peri-pherals	GPS	Network	Image processing	GIS	Database
	MF	WS	PC						
National Mapping and Resource Information Authority, Philippines		✓	✓	Scanner, digitizers, printers, film writer	Receivers and base stations		DISIMP, ERDAS, ER Mapper, micro-BRIAN	Arc Info, Mapinfo, SPANS	dBase, Oracle
National Water Resources Board, Philippines			✓	Digitizers, printers		LAN		Microstat	Foxpro, Oracle
U.P. Training Centre for Applied Geodesy and Photogrammetry, Philippines		✓	✓	Scanner, digitizers, printers	Receivers and base stations	LAN, WAN, Internet	DISIMP, ERDAS, ER Mapper, micro-BRIAN	Arc Info, Mapinfo, SPANS	
Systems Engineering Research Institute, Republic of Korea	✓	✓	✓	Scanner, digitizers, printers	Receivers	LAN	ERDAS, PCI	Arc Info, Idrisi	Oracle
Centre for Remote Imaging, Sensing and Processing, Singapore		✓	✓	Tektronix Phaser 480		LAN, Internet	ERDAS, PV-Wave, Vista/Prism	Arc Info	Ingress
Mahaweli Authority, Sri Lanka		✓	✓	Digitizers, printers	Receivers and base stations	LAN	ERDAS, PCI	Arc Info, Idrisi, SPANS	dBase
Thailand Remote Sensing Centre, Thailand		✓	✓	Scanner, digitizers, printers	Receivers	LAN, Internet	I²S, Meridian, TNTmips, PCI	Arc Info, ILWIS, SPANS	Oracle
National Centre for Science and Technology, Viet Nam			✓	Digitizers		LAN	ERDAS, ER Mapper, Pericolor	Arc Info, Idrisi, ILWIS, Mapinfo	dBase, Foxpro

Notes: MF, WS and PC stand for mainframe, workstation and personal computer, respectively.

Table 4.3 Estimated investment (US dollars) in infrastructure

Country	Agency	1992-1996	1996-2000
Australia	CSIRO Office of Space Sciences and Applications	10,000,000	10,000,000
Azerbaijan	Azerbaijan National Space Agency	40,000	200,000
China	Wuhan Technological University	110,000	–
China	National Remote Sensing Centre, Changsha	95,000	120,000
Fiji	Ministry of Lands, Mining and Energy	500,000	250,000
Fiji	University of South Pacific	130,000	200,000
Indonesia	National Aeronautics and Space Institute	6,500,000	8,500,000
Islamic Republic of Iran	Iranian Remote Sensing Centre	350,000	2,000,000
Japan	National Space Development Agency of Japan	14,000,000	5,000,000
Japan	Ministry of International Trade and Industry	1,000,000	500,000
Pakistan	Pakistan Space and Upper Atmosphere Research Commission	125,000	250,000
Philippines	Cybersoft Integrated Geoinformatics, Inc.	480,000	280,000
Philippines	National Water Resources Board	28,500	–
Philippines	University of the Philippines Training Centre for Applied Geodesy and Photogrammetry	85,000	250,000
Philippines	Geodata System Technologies, Inc.	205,000	550,000
Philippines	National Mapping and Resource Information Authority	2,500,000	–
Republic of Korea	Systems Engineering Research Institute	400,000	700,000
Singapore	Centre for Remote Imaging, Sensing and Processing	8,000,000	4,000,000
Sri Lanka	Mahaweli Authority	550,000	150,000
Thailand	Thailand Remote Sensing Centre	500,000	500,000

imaged by optical sensors with the desired frequency. Further, some satellites, such as SPOT and IRS-C, have the ability to obliquely image the ground at various "look" angles, affording stereoscopic views that are useful for topographic studies. While certain satellites/sensors are more oriented towards imaging the land, some others are designed for the study of the oceans. The former group includes Landsat, SPOT, IRS, the Russian Federation's Resurs and the AVNIR sensor on ADEOS, while the latter group includes Japan's MOS, the Russian Federation's Okean, the Topex/Poseidon, ERS and the OCTS sensor on ADEOS. The large number of operational remote sensing satellites currently in orbit means, in effect, a higher overall repetitiveness.

It is evident from the above that the spacefaring countries of the region have contributed, in a substantial way, to the development of satellite remote sensing technology by designing and launching a number of operational remote sensing satellites. These countries are expected to continue to play a significant global role in the development of the remote sensing space segment. For instance, Japan is developing ADEOS-2 and ALOS, as well as developing, jointly with the United States, the TRRM, specifically designed for tropical rainfall measurements and scheduled for launch in 1997. The Russian Federation plans to launch more satellites in its Resurs and Okean series for land and ocean observations, respectively, over the next few years. Resurs data are also being received at the ESA ground station at Kiruna, Sweden; the Kiruna station also markets the data. India will be launching more satellites in the IRS series, carrying various combinations of sensors. IRS data are also being received in the United States and are being marketed by EOSAT of the United States outside India. China is developing a dedicated remote sensing satellite jointly with Brazil.

4.1.3 GIS

The last few years have witnessed the emergence of GIS as a powerful tool with the versatility and flexibility to handle multiple layers of data from different sources in a spatially referenced framework. Since satellite remote sensing data are basically spatial data in digital form, they can be readily stored, integrated, manipulated and analysed with other spatial data sets in a GIS. There has thus been in recent years a tremendous upsurge in the combined or integrated use of these two versatile technologies in resource and environmental surveying, mapping and management activities. An increasing number of government and semi-government organizations, research institutions, universities, NGOs and private agencies in the Asian and Pacific region, representing a broad spectrum of disciplines and user interests, are involved in remote sensing and GIS applications.

4.1.4 Satellite-based positioning

Satellite-based positioning, such as the United States' NAVSTAR Global Positioning System, is also emerging as a highly useful tool that provides precise information on the position of any point on the Earth's surface in terms of geodetic and geographic coordinates and elevation. The user's portable receiver catches signals from a constellation of 24 satellites of the NAVSTAR GPS orbiting at an altitude of about 20,000 km, and converts them into position and navigation information. Very high accuracy is offered to the military, while civilians are provided with relatively lower accuracy. However, using differential GPS, civilian users can improve accuracy appreciably. The Russian Federation also has its own satellite-based system for navigation and positioning called GLONASS. With its ability to determine precisely the geographical coordinates of any place in a shorter time than conventional surveying methods, satellite-based positioning can be a very useful tool for mapping and can effectively supplement and complement remote sensing and GIS technologies in surveying and mapping projects. Satellite-based positioning is also equally useful for navigation and geodesy. It is thus being increasingly used in a variety of applications all over the world, including the Asian and Pacific region.

4.1.5 Conclusion

The region, as a whole, has developed substantial capabilities in the allied fields of satellite remote sensing and GIS. Most countries have developed a certain minimum level of facilities and expertise relating to remote sensing and GIS applications, while several have kept abreast with the continuing advances in the field. Some countries would be considered as being almost at par with developed countries of the West in this field and have contributed appreciably to both the space and the ground segments. The impressive range of remote sensing satellite systems designed and operated by regional countries, the large number of ground receiving stations functioning in the region and the dedicated facilities established for remote sensing and GIS applications by many countries would serve to show that the region has duly recognized the tremendous potential of remote sensing and is fully committed to using this technology in resource and environmental management. In some countries, remote sensing is being used in a more or less operational manner for some applications, while other countries are trying to move ahead from the present stage – where they essentially use remote sensing data for research, demonstration and pilot studies – towards more practical and operational use, as an integral part of their socio-economic development activities. The RESAP Regional Working Group on Remote Sensing, GIS and Satellite-based Positioning will, it is hoped, make some contribution to this effort. At its third meeting, held in May 1997, at Taejon, Republic of Korea, the Regional Working Group formulated a medium-term plan, based on a minimum common programme focusing on six priority themes or areas, namely forest resources monitoring, coastal zone management, desertification monitoring, urban and regional development, spatial information infrastructure and crop production forecasting. It is believed that the collaborative projects and activities to be carried out in these thrust areas would help participating countries in the operational application of remote sensing, GIS and satellite-based positioning.

With the region's substantial and continuously expanding capabilities in remote sensing and associated GIS and satellite-based positioning technologies, especially its commendable contribution to the remote sensing space segment, there is every hope that the region will continue to advance further in these fields and have even more significant effects on the world space technology scene. It is hoped that regional countries will move steadily towards operational use of remote sensing and allied technologies in support of the eventual goal of environmentally sound and sustainable development.

4.2 Country inputs

4.2.1 Azerbaijan

Azerbaijan National Space Agency

The Azerbaijan National Space Agency is a government organization whose mandate is specified in a decree issued by the President of Azerbaijan.

Professional Personnel

Primary field of work	Number of professional staff with practical experience in			Number of staff with the following qualifications			Number of staff with intensive training (>2 months) during last 5 years in		
	RS	GIS	GPS	Ph.D.	MSc	BSc	RS	GIS	GPS
Agriculture/forestry/rangeland management	16				4	12			
Ecology/biology/biodiversity	11				1	10			
Geology/geomorphology/ soil sciences	9				2	7			
Geography	13				2	11			
Meteorology/climatology	8			1	2	5			
Oceanography and fisheries	6				1	5			
Environmental monitoring									
Urban/regional planning									
Hydrology/water management	5				1	4			
Rangeland management									
Photogrammetry/cartography/ digital mapping	10					10			
Sensor development	10			2	2	6			
Hardware/software development									
Data management/information systems services									

Total number of professional staff: 88   Total number of administrative staff:

Major Projects and Activities (1992-1996)

Title	Geographical extent/surface area in km^2	Objectives	Generated products	Staff time	Names of other agencies involved in project/activity
Land-use inventory	66,000	Agricultural land square assessment	Maps and tables	960 person-years	Land State Committee
Flood zones in Caspian seashore territory of Azerbaijan	20,000	Determination of ecologically hazardous regions	Maps	240 person-years	State Committee on Hydrometeorology

Major Current/Ongoing Projects/Activities

Title	Geographical extent/surface area in km^2	Objectives	Generated products	Staff time	Names of other agencies involved in project/activity
Development of hardware/software complex for processing and automatic cartographic presentation	100,000	Creation of thematic maps	Thematic maps	25 person-years	
Creation of onboard measure-information system (MIS) for remote sensing airborne laboratory	100,000	Creation of system for real-time monitoring of underlying surfaces	On board MIS	50 person-years	
Creation of hardware/software for operative distribution of remote sensing data by ground communication channels	100,000	Creation of system for operative distribution of remote sensing data to users	Hardware/software complex	40 person-years	

Seminars/Conferences/Workshops Organized (1992-1996)

Type (seminar, conference, workshop, etc.)	Theme/title	Co-sponsorship (if any)	Number of participants	Title of output (publication and/or other)
International conference	Ecology problems		100	Transactions of the conference (Baku, 1994)

Training Activities (1992-1996)

Course title and description	Training undergone by organization staff		
	Given by (name of institution/agency)	Duration/frequency	Number of staff taking course
(none)			

Course title and description	Training provided by organization for external users		
	Name of participating agencies/organizations	Duration/frequency	Number of participants
Earth study from space; aerospace device development, measure-informative system	Azerbaijan State Oil Academy	Annual	50
Automatic systems for information processing and control	State Technology University	–	20

Databases/Archives Developed and/or Maintained and Available to Outside Users

Type (digital or analogue) and storage medium	Database contents (type, location, source data, scale and other useful information)	Geographical extent/surface area in km^2	Year of production/ review	Internet address (URL)	Main users
Analogue	Data on temperature, thermo and mass exhange, wind speed, waves, Bowen numbers for Azerbaijan sector of Caspian Sea; format A4, 300 maps	20,000	1995		State Oil Company of Azerbaijan
Analogue	General atlas of Absheron peninsula. Data on distribution of heavy metals, oil polluted regions, ecological state of plants, industrial zones, Baku and Sumgait; format A4, scale 1:25,100 maps	30,000	1996		

Documents Published since 1990

Title	Year of publication	Language of publication	Keywords	Frequency
Transactions of Conference on Ecology Problems, Baku	1994	English	Remote sensing, monitoring, ecology, aerospace information processing	
Space Science and Technology Symposium, Turkey	1993	English	Remote sensing, sensors, processors	

4.2.2 China

National Remote Sensing Centre, Changsha Branch

The National Remote Sensing Centre of China (Changsha Branch) is a government organization mandated to organize studies on policies of remote sensing development and related technological development, draw up long-term plans for remote sensing development, coordinate remote sensing technical forces in the country, identify priority development projects and organize national key projects relevant to economic construction, promote scientific and technological exchanges in remote sensing at both national and international levels, promote remote sensing research, development and applications, and provide remote sensing technical training and consultation services. The organization chart of this organization is shown in figure 4.1.

Professional Personnel

Primary field of work	Number of professional staff with practical experience in			Number of staff with the following qualifications			Number of staff with intensive training (>2 months) during last 5 years in		
	RS	GIS	GPS	Ph.D.	MSc	BSc	RS	GIS	GPS
Agriculture/forestry/rangeland management									
Ecology/biology/biodiversity									
Geology/geomorphology/soil sciences	12	2		2		12	3		
Geography									
Meteorology/climatology									
Oceanography and fisheries									
Environmental monitoring	8	4				8	2		
Urban/regional planning									
Hydrology/water management	5							2	
Rangeland management									
Photogrammetry/cartography/digital mapping	1					1			
Sensor development				2					
Hardware/software development	1					1			
Data management/information systems services	1					1			

Total number of professional staff: 23   Total number of administrative staff:

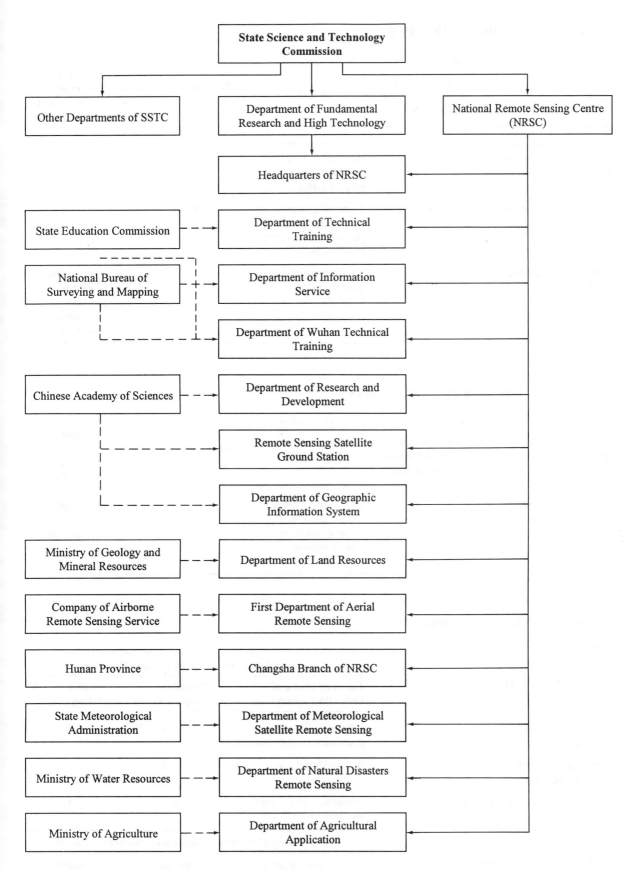

Figure 4.1 Organization chart of the Chinese National Remote Sensing Centre

C

China

Major Projects/Activities (1992-1996)

Title	Geographical extent/surface area in km²	Objectives	Generated products	Staff time	Names of other agencies involved in project/activity
Remote sensing/geological exploration in southern Hunan	45,000	To study geological structure and features for identifying possible mineralization zones	Maps		
Database for environmental monitoring	Four cities	To build up database on environmental parameters, including groundwater	Database		
GIS of Dongting Lake in Hunan		To minimize the disastrous effects of floods			Geographic Institute of Chinese Academy of Sciences

Major Current/Ongoing Projects/Activities

Title	Geographical extent/surface area in km²	Objectives	Generated products	Staff time	Names of other agencies involved in project/activity
Geological information system of Human		To help in mineral exploration by identifying areas likely to have minerals			Regional Geological Institute

Training Activities (1992-1996)

Course title and description	Training undergone by organization staff		
	Given by (name of institution/agency)	Duration/ frequency	Number of staff taking course
Remote sensing and GIS training	Official development assistance (United Kingdom)	4 months	1
Computer and remote sensing imaging processing	Beijing University	2 months	3
Environmental monitoring database	Department of Geology and Minerals	2 weeks	1

Course title and description	Training provided by organization for external users		
	Name of participating agencies/organizations	Duration/ frequency	Number of participants
Remote sensing image processing and GIS training	Remote Sensing Station	1 month	28

Documents Published since 1990

Title	Year of publication	Language of publication	Keywords	Frequency
Index Atlas of GIS of China	1990	English		
Bulletin of Flood Damage Evaluation Information System and Wetland Use	1991	English		

C

Wuhan Technical University of Surveying and Mapping

Professional Personnel

Primary field of work	Number of professional staff with practical experience in			Number of staff with the following qualifications			Number of staff with intensive training (>2 months) during last 5 years in		
	RS	GIS	GPS	Ph.D.	MSc	BSc	RS	GIS	GPS
Agriculture/forestry/rangeland management	2	2	3	3	3	1	2	2	3
Ecology/biology/biodiversity									
Geology/geomorphology/ soil sciences									
Geography		1				1			
Meteorology/climatology	2		1		2	1	2		1
Oceanography and fisheries	2		2		2	1	2		1
Environmental monitoring	2	2		2	2		2	2	
Urban/regional planning		5	2	2	5			5	2
Hydrology/water management									
Rangeland management	2	5	2	2	5	2	2	5	2
Photogrammetry/cartography/ digital mapping	4	10	2	5	11		4	10	2
Sensor development	2				2		2		
Hardware/software development	5	20	10	10	15	10	5	20	10
Data management/information systems services									

Total number of professional staff: Total number of administrative staff:

Major Projects/Activities (1992-1996)

Title	Geographical extent/surface area in km^2	Objectives	Generated products	Staff time	Names of other agencies involved in project/activity
GPS framework	Whole country	For primary control for mapping			National Bureau of Surveying and Mapping
Urban GPS network	Over 20 cities	For urban mapping control			

Title	Geographical extent/surface area in km²	Objectives	Generated products	Staff time	Names of other agencies involved in project/activity
GPS monitoring network	Qingzhang Plateau	To monitor crustal movement			National Bureau of Surveying and Mapping
GPS data processing software			Commercial software		
GIS project	Over 10 cities	To assist in land management, urban planning, municipal infrastructure management			
GIS software development		To develop commercial software			
Remote sensing applications	Over 10 cities and counties	To assist in environmental monitoring, urban planning, agriculture, regional planning			
Digital photogrammetry workstation		For softcopy photogrammetry			

Major Current/Ongoing Projects/Activities

Title	Geographical extent/surface area in km²	Objectives	Generated products	Staff time	Names of other agencies involved in project/activity
GPS monitoring for safety of Qinjiang Dam	Qinjiang Dam Reservoir	To develop automatic monitoring system for the safety of Qinjiang Dam			
GIS software development					
Digital photogrammetry workstation					

Training Activities (1992-1996)

Course title and description	Training undergone by organization staff		
	Given by (name of institution/agency)	Duration/ frequency	Number of staff taking course
(none)			

Course title and description	Training provided by organization for external users		
	Name of participating agencies/organizations	Duration/ frequency	Number of participants
"3S" integration seminar	Concerned agencies	1-2 months a year	About 60

Department of Remote Sensing of Forestry, Chinese Academy of Sciences

The Department of Remote Sensing of Forestry of the Chinese Academy of Sciences is a government organization mandated to take the lead and coordinate all activities regarding remote sensing of forestry at the national level.

Professional Personnel

Primary field of work	Number of professional staff with practical experience in			Number of staff with the following qualifications			Number of staff with intensive training (>2 months) during last 5 years in		
	RS	GIS	GPS	Ph.D.	MSc	BSc	RS	GIS	GPS
Agriculture/forestry/rangeland management	4			1	3				
Ecology/biology/biodiversity	1				1				
Geology/geomorphology/ soil sciences	1				1				
Geography									
Meteorology/climatology									
Oceanography and fisheries									
Environmental monitoring	1					1			
Urban/regional planning	1				1				
Hydrology/water management									
Rangeland management	1				1				
Photogrammetry/cartography/ digital mapping	1					1			
Sensor development									
Hardware/software development	3				2	1			
Data management/information systems services	1					1			

Total number of professional staff: 14   Total number of administrative staff:

Major Projects/Activities (1992-1996)

Title	Geographical extent/surface area in km²	Objectives	Generated products	Staff time	Names of other agencies involved in project/activity
Monitoring and assessment of forest fires using remote sensing	900,000	To develop an operational dynamic monitoring system			Beijing Forestry University Beijing Normal University
Application of SAR in forestry	100,000				
Monitoring land use in northern China	2,200,000				

Major Current/Ongoing Projects/Activities

Title	Geographical extent/surface area in km²	Objectives	Generated products	Staff time	Names of other agencies involved in project/activity
Application of SAR in monitoring paddy				40 person-years	Beijing University
Establishment of environmental and resource information system	600,000	To develop an information system for the Ministry of Forestry		30 person-years	Institute of Remote Sensing, Chinese Academy of Science
Perfecting the forest fire monitoring system		To develop a commercial monitoring system			

Seminars/Conferences/Workshops Organized (1992-1996)

Type (seminar, conference, workshop, etc.)	Theme/title	Co-sponsorship (if any)	Number of participants	Title of output (publication and/or other)
Seminar	Remote sensing of tropical rain forest in China	Joint Research Commission, European Union	20	

Databases/Archives (Including Remotely-Sensed Data Developed and/or Maintained and Available to Outside Users

Type (digital or analogue) and storage medium	Database contents (type, location, source data, scale and other useful information)	Geographical extent/surface area in km²	Year of production/ review	Internet address (URL)	Main users
Analogue	Dynamic map of forest coverage in Hebei Province	200,000	1992		Ministry of Forestry
Analogue	Land-use map of northern China	2,300,000	1995		Relevant departments
Analogue	Hazards of forest fires in south-west China	1,100,000	1995		Forestry Bureau in south-west China
Digital, tape	Simulation of forest fire spread	100,000	1994		Scientific agencies

Documents Published since 1990

Title	Year of publication	Language of publication	Keywords	Frequency
Proceedings: Monitoring and Assessment of Remote Sensing on Forest Fires	1995	Chinese		
Proceedings: The Application of SAR to Forestry	1996	Chinese		
150 articles, papers, etc., relating to remote sensing/GIS applications to forestry	1991-1995	Chinese		

4.2.3 Fiji

Ministry of Lands, Mining and Energy

The Ministry of Lands, Mining and Energy is a government organization mandated to provide expertise in remote sensing and geographic information system to enable the government, public and agencies to improve the use and management of resources and assets.

Professional Personnel

Primary field of work	Number of professional staff with practical experience in			Number of staff with the following qualifications			Number of staff with intensive training (>2 months) during last 5 years in		
	RS	GIS	GPS	Ph.D.	MSc	BSc	RS	GIS	GPS
Agriculture/forestry/rangeland management									
Ecology/biology/biodiversity									
Geology/geomorphology/ soil sciences	1				1		2		
Geography									
Meteorology/climatology									
Oceanography and fisheries									
Environmental monitoring									
Urban/regional planning	1	2							2
Hydrology/water management									
Rangeland management									
Photogrammetry/cartography/ digital mapping	10				1		5		
Sensor development									
Hardware/software development	2				2				
Data management/information systems services	4						4		

Total number of professional staff: Total number of administrative staff: 26

Major Projects/Activities (1992-1996)

Title	Geographical extent/surface area in km²	Objectives	Generated products	Staff time	Names of other agencies involved in project/activity
Fiji land information system (FLIS)	Fiji islands and reef area	To computerize the "core" land-related data	Digital Hard copy		Registrar of Titles, Director of Town and Country Planning, Native Land Trust Board

Fiji

Title	Geographical extent/surface area in km²	Objectives	Generated products	Staff time	Names of other agencies involved in project/activity
Fiji topographic database	Fiji islands and reef area	To establish a complete digital topographic database and to accelerate production of 1:50,000 scale topographic maps	Digital Hard copy		New Zealand Government
Census mapping	Fiji islands	To provide E.A. boundary maps	Digital Hard copy		Bureau of Statistics, UNDP

Major Current/Ongoing Projects/Activities

Title	Geographical extent/surface area in km²	Objectives	Generated products	Staff time	Names of other agencies involved in project/activity
Fiji digital topographical database	Fiji Island	To establish a national digital topographic database to accelerate the production of 1:50,000 topographic series	Digital Hard copy		Terrallink, New Zealand
Native land mapping system	Fiji Island	To integrate native land ownership data with cadastral maps to have complete mapping of land holdings in Fiji	Digital Hard copy		Native Land Trust Board
Fiji islands 1:50,000 topographic mapping	Fiji Island		Hard copy		

Seminars/Conferences/Workshops Organized (1992-1996)

Type (seminar, conference, workshop, etc.)	Theme/title	Co-sponsorship (if any)	Number of participants	Title of output (publication and/or other)
Workshop	Remote sensing and GIS for land and natural resources and environment management in the Pacific subregion	ESCAP	72	Proceedings of workshop
Seminar	Fifth regional remote sensing seminar on tropical ecosystem management	ESCAP and NASDA	77	Proceedings of seminar

Training Activities (1992-1996)

Course title and description	Training undergone by organization staff		
	Given by (name of institution/agency)	Duration/ frequency	Number of staff taking course
Regional remote sensing seminar on tropical eco-system management	NASDA and RESTEC	3 days	5
Introduction to map information	SORC, Fiji	5 days	20
Introduction to GIS	University of Queensland, Australia	3 months	3

Databases/Archives Developed and/or Maintained and Available to Outside Users

Type (digital or analogue) and storage medium	Database contents (type, location, source data, scale and other useful information)	Geographical extent/surface area in km^2	Year of production/ review	Internet address (URL)	Main users
Digital	Survey records, valuation assessment records, native cadastral mapping	Fiji islands	1992		Government and private sector
Analogue	Lease administration, native land ownership records, native land mapping index, topographic maps	Fiji islands	1992		Government and private sector
Analogue	Fiji geodetic records, census mapping, town and country planning, application system	Fiji islands	1992		Government and private sector

Documents Published since 1990

Title	Year of publication	Language of publication	Keywords	Frequency
A Coordinated Approach to Land Information Management		English	Fiji land information system	Every quarter
FLIS News		English		Every quarter
FLIS Charging Policy		English	Charging for land information	

University of the South Pacific

The University of the South Pacific is an educational institution mandated to provide, *inter alia*, education in remote sensing and GIS to both students and professionals, engage in research and services, and promote the development of these technologies in the South Pacific region.

Professional Personnel

Primary field of work	Number of professional staff with practical experience in			Number of staff with the following qualifications			Number of staff with intensive training (>2 months) during last 5 years in		
	RS	GIS	GPS	Ph.D.	MSc	BSc	RS	GIS	GPS
Agriculture/forestry/rangeland management	2	2		2		1			
Ecology/biology/biodiversity	2	2		2		1			
Geology/geomorphology/ soil sciences	3	1		3					
Geography	4	2		4		1			
Meteorology/climatology									
Oceanography and fisheries									
Environmental monitoring	3	2		3		1			
Urban/regional planning	1	1		1					
Hydrology/water management	2	1		2					
Rangeland management									
Photogrammetry/cartography/ digital mapping	2	2		2					
Sensor development									
Hardware/software development		1		1					
Data management/information systems services									

Total number of professional staff: 4   Total number of administrative staff: 2

Training Activities (1992-1996)

Course title and description	Training undergone by organization staff		
	Given by (name of institution/agency)	Duration/ frequency	Number of staff taking course
Introduction to GIS	University of the South Pacific	1 to 2 per year	45
Advanced GIS	University of the South Pacific	1 per year	30
Remote sensing	University of the South Pacific	1 per year	30

4.2.4 India

FACILITIES AND ACTIVITIES*

Introduction

During the last three decades, developments in space technology and its applications have made great strides and contributed significantly to key sectors of development, such as natural resources management, disaster management, environmental monitoring, meteorology, telecommunication, television and radio broadcasting and education. In India, the two space systems, namely, the Indian Remote Sensing Satellite System (IRS) and the Indian National Satellite System (INSAT) are providing operational services in these key sectors of development.

Recognizing the importance of natural resources management, the Planning Commission, Government of India, set up the National Natural Resources Management System (NNRMS) in 1983. NNRMS is conceived as a system meant to facilitate optimal utilization of the country's natural resources through a proper and systematic inventory of the resource availability as well as to reduce regional imbalances through effective planning. The Department of Space is the nodal agency for establishing NNRMS. To guide the evolution of such a system, a preparatory committee (later renamed Planning Committee PC-NNRMS) was constituted by the Planning Commission. Eight task forces set up by PC-NNRMS with experts from various resource sectors conducted an in-depth study on the technical suitability, cost effectiveness, accuracy and complementary use of remotely-sensed data with conventional surveys and techniques.

Based on the recommendations and suggestions by the task forces, six standing committees were set up for the sectors of (a) agriculture and soils, (b) bio-resources and the environment, (c) geology and mineral resources, (d) ocean resources, (e) water resources and (f) remote sensing technology and training. The standing committees are chaired by Secretaries of the respective departments who are assisted by experts from major user agencies. The Committees provide guidelines on major issues related to the themes, identify new areas for research and development and advise on specific national programmes.

Concurrently, the experience gained in conceptualizing and implementing the space segment with the necessary ground data reception, processing and interpretation system and in integrating the satellite remote sensing data with conventional data systems for resource management, paved the way for initiating an application-driven programme in the country.

With the successful completion of IRS-UP projects, some of the major applications have been identified in the fields of agriculture, soils and land use, water resources, ocean and marine resources, forestry and geology to be taken up in "mission mode". Large-scale operational demonstrations are carried out in these selected themes under the umbrella of a programme called Remote Sensing Application Missions (RSAM). These missions are closely monitored by a management council in order to expedite the transfer of these theme projects to the users, once they are operational. The noteworthy aspect of these RSAM projects is that many user departments have accepted the utility of remote sensing for implementing their management plans and hence have been funding a number of national-level application projects.

With active participation and funding support from user departments and agencies in both the central and state governments, remote sensing technology has been operationalized to cover diverse resource themes and areas such as forestry, agricultural crop acreage and yield estimation, drought monitoring and assessment, flood monitoring and damage assessment, land-use and land-cover mapping, wasteland management, water resources management, groundwater targeting, fishery potential forecasting, urban planning, mineral targeting, and environmental impact assessment.

---

* Based on a report provided by the Indian Space Research Organization.

The major emphasis of NNRMS has been the conducting of application studies as per the user requirements, establishment of necessary infrastructure for remote sensing at various levels, technology development, generation of trained manpower, etc.

A. Status of remote sensing infrastructure facilities

To operationalize NNRMS in the country, the Department of Space (DOS) has been making concerted efforts in setting up remote sensing infrastructure facilities at different levels viz., regional, state and line departments.

1. Facilities with national organizations

Remote sensing and/or geographic information system facilities are available with various central government organizations such as the Geological Survey of India, Atomic Minerals Division, Oil and Natural Gas Corporation, Ltd., Mineral Exploration Corporation, Ltd., Central Ground Water Board, National Institute of Hydrology, Central Water Commission, National Bureau of Soil Survey and Land-use Planning, All India Soils and Land-use Planning, Forest Survey of India, National Institute of Oceanography etc. In addition, DOS centres and units, namely the National Remote Sensing Agency, Space Applications Centre and Indian Space Research Organization Headquarters, have adequate facilities to carry out remote sensing and GIS-related projects as well as conducting training programmes.

2. Regional remote sensing service centres

Presently five regional remote sensing service centres (RRSSC) established under the umbrella of NNRMS at Bangalore (southern region), Nagpur (central region), Jodhpur (north-western region), Dehra Dun (northern region) and Khargpur (eastern and north-eastern region) are providing regional-level services and training on digital analysis of satellite data.

3. State remote sensing applications centres

Towards establishing remote sensing infrastructure at state level, DOS has been continuously interacting with state governments (with the Chief Secretary and Secretary of Science and Technology and Planning) as well as with the Planning Commission. As a result of these efforts by DOS, 23 states (excluding Delhi, Meghalaya and Tripura) have already established remote sensing applications centres/units and are carrying out remote sensing-related projects. DOS has been providing not only technical guidance but also grant-in-aid for establishing or strengthening such state remote sensing applications centres. At the suggestion of DOS, the Planning Commission has also created a separate sub-head for fund allocation for remote sensing applications centres in the state annual plan budget under the space and technology sector.

The state remote sensing centres are equipped with suitable infrastructure for visual interpretation, most of them with digital image processing systems and also GIS. Some of the state centres, like the ones in Uttar Pradesh, Tamil Nadu, Punjab and Haryana, are equipped with a main frame image analysis system. In order to effectively utilize the IRS-1C stereopan data for generation of DEM and cartographic applications, the state centres of Tamil Nadu, Andhra Pradesh, Punjab, Orissa and Maharashtra are in the process of procuring digital stereo-analysis systems.

Most of the centres are actively participating in the national missions being coordinated by DOS such as nationwide land-use and land-cover mapping for agro-climatic zones planning, nationwide wasteland mapping on a 1:50,000 scale, nationwide wetland mapping on a 1:250,000 scale, crop acreage and production estimation, cotton acreage and condition assessment, national river action plan for sewerage treatment plants, coastal zone regulation mapping, and integrated mission for sustainable development. These centres are also carrying out various projects that are sponsored by central government departments and agencies, state government departments and agencies and private agencies. Most of the state remote sensing centres also conduct training programmes or workshops for the benefit of resources scientists in various state government departments.

In addition to the state remote sensing applications centre, some states have set up facilities for remote sensing-related activities in line departments concerned with various fields, such as geology, ground water, soils and agriculture etc.

4. Academic institutes

Academic institutes like the Indian Institute of Technology (at Bombay, Khargpur, Delhi), Roorkee University, Anna University, Andhra University, Aligarh Muslim University, Jawaharlal Nehru Technological University, Bharathidasan University, Sagar University, Indian Agricultural Research Institute, and others have remote sensing and/or GIS facilities to aid in educational programmes and in carrying out R and D projects.

B.   Details of the regional remote sensing service centres

A summary of the activities and facilities of the five regional remote sensing centres established by DOS during the period 1986-1987 is given below:

1. Location

Bangalore, Dehra Dun, Nagpur, Khargpur and Jodhpur.

2. Status of the Department

Status:

(a)   Parent department Department of Space;
(b)   Autonomous status.

3. Name of the Head of the centre

Shri A Adiga – Bangalore
Shr M L Manchanda – Dehradun
Mr Y N Krishnamurthy – Nagpur
Shr S Sudhakar – Kharagpur
Mr J R Sharma – Jodhpur

4.   Addresses

Regional Remote Sensing Service Telephone/Telex Centre
40th Main, Eshwarnagar, Behind PSTI
Bangalore – 560 070
Tel: 080-6633795
Fax: 0135-745439

Regional Remote Sensing Service Centre
IIRS Campus, P.B. No. 135
4, Kalidas Marg
Dehra Dun – 248 001
Tel: 0135-742439
Fax: 0135-745439

Regional Remote Sensing Service Centre
P.O. No. 439, Shankar Nagar P.O.
Nagpur – 440 010
Tel: 0721-531339
Fax: 0721-547408

Regional Remote Sensing Service Centre
IIT Campus, Kharagpur – 721 302
Tel: 03222-55466
Fax: 03222-77201

Regional Remote Sensing Service Centre
CAZRI Campus, Jodhpur – 342 003
Tel: 0291-40004
Fax: 0291-41516

5. Facilities and infrastructure

All the centres are provided with identical digital analysis, photo processing and visual interpretation facilities. Given below are the composition of each facility:

(a) Digital analysis facility

(i) VAX-11/780-based image processing system: this system is configured around a VAX-11/780 super minicomputer, augmented with an array processor and 3 Numelec PC-2001 image workstations. The system comprising two tape drives, 2.4 GB disk system, printer, plotter, digitizer and a Dunn camera, is integrated with a powerful VIPS-32 image processing package and IPRLIB image processing library.

(ii) Workstation-based image analysis system: the image processing and GIS facilities of the Centres have been augmented with the addition of eight IBM RS/6000 Model 37T workstations with EASI/PACE image analysis software package and Arc Info GIS package.

Configuration details:

- 62 MHz power CPU
- 32/64MB RAM (expandable to 256 MB)
- 8 GB Dual access disk drive
- 1 GB SCSI-II disk
- Integrated SCSI-II interface
- 3.5", 1.44 MB floppy drive
- Additional SCSI-II interface
- Power Gt4i graphics adapter
- 20" colour monitor (1280 x 1024 pixels)
- Integrated ethernet interface
- Keyboard and mouse
- 2 Serial 1 parallel port.

(iii) In-house software library: in addition, a comprehensive library of software modules developed by the Centres are available. This includes specific packages for application missions like land-use/land-cover mapping, crop acreage and production estimation (CAPE), vegetation index map generation using NOAA data, change detection and a large number of utilities to meet specific application requirements.

(b) Geographic information system

- Centres are equipped with Arc Info Version 7.0 on IBM RS-6000 workstations.
- Indigenously developed GIS packages viz. PC-INGIS, INGIS and GEOSPACE

(c) Micro-photogrammetry system

The Centres at Bangalore and Kharagpur are equipped with a micro-photogrammetry system (MPS-II). This system facilitates aerial triangulation and three-dimensional analysis for topographic mapping.

(d) *Photowrite system*

Each centre is also equipped with a colour photowrite system which is a digital film recorder for producing continuous tone B and W or colour films directly from digital image data. This facilitates generation of photographic outputs of processed image data of any size in the desired scale.

(e) *Full-scale scanner*

The centre has a CONTEX FSS 8000 scanner with software having scanning, viewing, printing and data format conversion facilities. The scanner runs on various workstations.

(f) *Photo processing facility*

- Klimsch colorotronics
- Colenta colour processor
- Kinderman development system

(g) *Visual interpretation and ground truth equipment*

- Hand-held global positioning system
- Procom-2
- Light tables
- Mirror stereoscopes
- NC scriber
- Spectroradiometers

(h) *Support system*

- Uninterrupted power supply (55 KVA capacity)
- Air-conditioning plat (30 T)
- Diesel generator (150 KVA)

(i) *Library*

Each Centre houses a functional library containing essential books/journals relating to remote sensing.

6. Proposed facilities

- Memory upgrading
- Network upgrading
- Server upgrading
- Disk upgrading

7. Manpower

Each Centre has the minimal specialized manpower to meet the operational requirements of the Centre. These may be categorized as follows: Head of the centre, thematic specialists, computer specialists, photography specialists, tradesmen, draughtsman, administration and support:

Head:	1
Thematic specialists:	5
Computer specialists:	6
Administration and support:	9

8. Projects completed/on-going with the details on salient achievements

(a) *Mission projects*

Some of the projects executed at the Centres are in mission mode. These aim at operational demonstration of remote sensing in resource management and development.

(i) Project Vasundhara

Application of remote sensing in geology is well established. The patterns of faults and fractures are analysed to show any regional stress directions and local structures such as concealed folds, mineral rock intrusive or major faults concealed in buried formations. The Geological Survey of India has taken up a joint project called Project Vasundhara with ISRO, Department of Space. This project envisages an integral appraisal of data from satellite remote sensing, airborne geophysics and ground geological, geophysical, geo-chemical data for mineral targeting on regional scale.

(ii) Generation of NDVI images for drought monitoring project

Drought monitoring at the all-India level has been executed by NRSA at RRSSCs for the year 1988, 1989 and 1990. Under the drought monitoring and early warning project, vegetation index maps, which are based on the concept that vegetation vigour is an indicator of drought, have been prepared viz. NOAA AVHRR data. The composite vegetation index maps are compared with ground parameters to establish interrelationship. The drought assessment bulletins from National Agricultural Drought Assessment Management System (NADAMS) are generated based on the analysis of both vegetation index map and the statistics. Total methodology development is done jointly by NRSA and RRSSC. The project execution was done by various RRSSCs.

(iii) Land-use mapping for agro-climatic zones

Inventory and mapping of land-use and land cover, using high-resolution data for both IRS-1A and Landsat have been completed for the whole country (level II) at 1:250,000 scale and mapping of culturable wasteland of specific areas at 1:50,000 scale. Towards this, a detailed methodology has been developed with the help of other agencies within and outside the Department of Space. The digital analysis of part of the projects was executed totally at the RRSSCs.

Realizing the need for up-to-date nationwide land-use and land-cover maps by several departments in the country, as a prelude a land-use and land-cover classification system (with 24 categories up to level II, suitable for mapping on 1:250,00 scale) was developed at NRSA, taking into consideration the existing land-use classification adopted by NATO, CAZRI, AIS and LUS, etc. and the details obtainable from satellite imagery. After detailed discussions with different user agencies, the classification system was modified and finalized. Based on the pilot studies, the Planning Commission has approved the nationwide land-use/land-cover mapping for agro-climatic zone planning. Operational methodology of digital analysis developed has been used for mapping 168 districts at different work centres.

(iv) Integrated resources mapping, Chandrapur

Consistent with the district-level planning requirement, thematic maps are generated on a 1:50,000 scale. Both the digital and visual techniques are followed interactively. Special techniques of stratification, layered approach, composition, aggregation and refinements are adopted wherever necessary to improve the quality of mapping. The primary thematic maps generated are geomorphology, geology, hydrology, soils, land use and land cover, forest/vegetation.

Basic maps are used to produce utilitarian types of maps to serve planning decisions. They are derived by a combination of two or more thematic maps or chosen parameters of the different themes. These are slope, land capability class, run-off potential, irrigability classes, groundwater potential.

(v) Integrated mission for sustainable development

Sustainable development of natural resources is based on maintaining the fragile balance between productivity functions and conservation practices through monitoring and identification of problem areas which require application of alternative agricultural practices, crop rotation, use of bio-fertilizers, energy-efficient farming methods and reclamation of underutilized lands.

All the RRSSCs are participating in this project. Using IRS-1B/P2 LISS-II data, various thematic maps are generated viz. land use/land cover, soil types, forest cover/types, groundwater prospects, surface water bodies, lithology/mineral occurrences, slope, etc.

These layers are integrated using an indigenously developed GIS package, and composite mapping units were generated and refined using logical as well as spatial grouping of the resource data. A specific watershed and specific action plans are being generated for implementation.

A total of 174 districts falling in various problem areas of the country such as drought prone areas, deserts, flood-prone areas, hill areas, etc. have been identified in consultation with the Planning Commission and respective state governments and taken up in mission project mode. Various centres of the Department of Space are collaborating with the concerned state remote sensing centres and academic/research institutions in carrying out this task. The management structure has been formulated and RRSSCs have initiated the work of the identified districts which are allocated to them.

(vi) Crop acreage and production estimation

The digital analysis of remote sensing satellite data for this project is mostly carried out at the RRSSCs. Digital analysis support has been providing for major crops, viz. rice, wheat, jowar, mustard, groundnut, oilseeds, sugarcane, sorghum, etc.

Digital analysis support to SAC and NNRMS for acreage, production and condition assessment for cotton and mulberry is provided at the RRSSCs.

(b) NRIS pilot projects

(i) GIS for development of micro-watershed

This project is initiated with the objective of creating a digital map for use along with the field parameters to arrive at a suitable model for generating a treatment plan for the micro-watershed in Kalyanakere watershed of Karnataka State. The Arc Info GIS package is being utilized for this purpose. All spatial data are integrated into the database. Derivation of specific thematic layers like run-off, slope, etc. are in progress and categorization criteria are under development.

(ii) GIS approach for mineral targeting in Gujarat

The objective of the project is to create a geological information system for the entire Gujarat State on 250-metre resolution and develop a geo-statical model for mineral prognostics. In phase I, the analysis is confined to Kutchh District. Base map preparation of this area is completed. Integration of geophysical and geo-chemical data is in progress. Broad parameters are shortlisted to implement the prognostic modelling of bauxite, one of the minerals identified.

(iii) Watershed prioritization

This project is aimed at prioritization of a micro-watershed through a GIS approach. The watersheds selected were Tawa reservoir and Indirasagar reservoir areas. All thematic layers, viz. land use, groundwater potential, slope, soil, geology, etc. are generated and integrated using INGIS. Composite mapping units have been generated and refined. Prioritization with reference to the above-mentioned two areas has been completed. Validation of results is in progress.

(c) Application validation projects and technology development programmes

Application validation projects are aimed at validating digital analysis methodologies for various applications with a view to standardization of application packages for operational use in resource management. These projects were evolved and validated jointly with actual user agencies. In addition, RRSSCs have undertaken the following studies with user agencies:

- Soil moisture studies
- Lithologic mapping
- Glacial studies
- Forestry-related studies

(d) User-funded application projects

Various user agencies have approached and executed many applications projects using digital and visual analysis facilities of RRSSCs. These projects are funded by the corresponding user agencies.

(e) Training programmes

Each RRSSC conducts training courses for scientific and technical personnel from central and state governments and academic R and D institutions in the area of digital image processing, geographic information system and applications of remote sensing for resource mapping and management. There are two different modules. One of the modules is of two weeks duration which is for working-level personnel. The courses involve lectures, practical exercises and demonstration on VAX-11/780 and IBM RS-6000. The other module is of 3-5 days and meant for senior officers.

(f) Software development activity

Even though the image analysis package (VIPS-32 system) has an exhaustive library of functions, in due course many requirements surfaced to develop software modules and packages for specific thematic applications.

(g) Ongoing projects

The RRSSCs have become the main driving force for processing remote sensing and conventional data in the country today. With the growing awareness of the use of remote sensing among the user community, there has been a rapid increase in user projects. This has also helped the Centres to become exposed to new areas of remote sensing applications.

9. Activities planned during ninth plan period

(a) Operational

(i) NRIS

Recognizing the need for a comprehensive information system essential for decision-making, NNRMS should develop a natural resources information system (NRIS) as a major component of NNRMS. The information system thus developed would provide periodic, updated and systematic information on land, water and ocean resources along with socio-economic data. Such an integrated database would help to plan the development and utilization of natural resources appropriately.

Under NRIS, a separate and dedicated computer system, GIS and manpower must be developed at each RRSSC. It would address specifically the needs of decision makers at three levels, namely the strategic, tactical and operational levels. NRIS should have a link with District Information System of National Informatics Centres (DISNIC). The information generated under various national missions, viz., IMSD, land use, wasteland and drinking water missions would serve as a base for NRIS.

All the state remote sensing centres, various central and state organizations, NGOs, academic and research institutions which are involved in resource mapping and collection and management of socio-economic data also should have close link with RRSSCs.

(ii) IMSD

One of the main operational thrust area for RRSSCs is going to be the implementation and monitoring of the IMSD programme. Action plan implementation needs to be closely monitored using satellite remote sensing, and corrective measures need to be incorporated for implementation as and when required.

(iii) Crop acreage and production estimation

Crop acreage and production estimation (CAPE) will continue to be operational in the future scenario. With availability of new sensors, additional crop types can be incorporated under this programme.

(iv) Value-added services

RRSSCs have already carried out a number of user projects in various fields like forestry, land use, soil, geology, hydrology, groundwater and so on. In addition, RRSSCs have developed expertise in handling national projects like IMSD. RRSSCs can take up similar projects from other departments of India and developing countries under value-added service. Major thrust areas are:

- Land and water resource development planning
- Forest monitoring, inventory and working plan
- Management of biodiversity

(v) Cadastral survey and mapping

The payload of IRS-1C scheduled for launch by the end of the 1995 would include a panchromatic band with 5.8 m resolution and 27 degree of nadir viewing capacity. Introduction of such high resolution with stereo-viewing capacity is going to be immensely useful to the community engaged in development and planning activities in urban and rural areas, as this would help in preparing and updating of large-scale maps up to 1:10,000 scale for urban/rural environments and mapping of resources and overlaying on cadastral-level maps at 1:10,000 or better scale.

(vi) National soil erosion inventory and monitoring using IRS data

Land degradation due to soil erosion constitutes the major portion of degraded lands. An inventory of eroded lands all over the country can be very much useful for prioritization and reclamation. Remote sensing techniques can be used for making an accurate inventory of soil erosion. A period check of eroded as well reclaimed areas can give an accurate account of these activities. IRS-1C with its very high resolution has a great potential for mapping eroded areas on a large scale and it is proposed to use these data for this project.

(vii) Integrated forest resources management system

Remote sensing and GIS can be used to develop an integrated forest resources management system for better management.

(viii) Land use and land cover

- Land-use and land-cover mapping of the entire country for 1995-1996 at 1:50,000 scale
- Change detection studies with respect to 1988-1989 land-use and land-cover mapping

Urban

- Urban land-use mapping at 1:10,000/25,000 scale
- Urban sprawl monitoring
- Urban land-use zonation

Agriculture and horticulture

- Total enumeration of different crops (district/taluk crop maps)
- Sample segment approach for statistics
- Online database for historical crop details
- Locust surveillance and control
- Desertification studies

Soil and land degradation

- Soil mapping at 1:50,000/25,000 scale

(ix) Forestry applications

Mapping and monitoring of wildlife sanctuaries and national parks is possible at regional level. Biodiversity and measurement of landscape elements like average patch size, patch diversity and patch interspersion can be done in the operational mode for some of the vast forest cover like the western Ghats and Himalayan areas. Near real-time forest fire monitoring and delineation of fire risk zone maps for forest fire prone areas can be done during the IRS-1C/1D mission period.

(b) Technology development projects

The ninth five-year plan will be a crucial transition period from the preceding years of research and development of air- and space-borne optical and microwave sensors to the routine application of satellite data. The integration of different data sets must overcome the problem of merging data with different spatial resolution on to map projections. With this, intercomparisons can be made and the full scientific value of the data realized.

To realize the scientific value of the data, the following projects have been proposed.

(i) Automated forest cover detection

Accurate detection of forest cover is a challenging job. However, experience shows that closed canopy forests are much easier to recognize from satellite images. We propose to develop methodologies by carrying out iterative research for automation of forest cover detection.

(ii) Preparation of forest working plan

Remote sensing technology has the potential to provide much valuable information for the generation of forest working plans.

(iii) Application of microwave data for ecosystem analysis

Application of microwave data in ecosystem analysis is at the developing stage. Microwave data are expected to assist in generating valuable information regarding canopy profile, under-storey, timber volume and biomass estimations. The proposed study will analyse all these parameters using multifrequency SAR (as available from SIR) data.

Thermal existence and energy balance modeling for forest canopies can be done using multi-spectral and TIR data.

(iv) Operational package for forestry applications

For better forest management and actual field applications of remote sensing and GIS, it is proposed to establish a forestry information system (FIS) initially at circle level and in future at division level. For the proper functioning of FIS, an operational package needs to be developed with a foolproof methodology.

(v) Integrated environmental management plan for the western Himalayas

Information on the Himalayan environment is scattered and often based on site-specific problems. An integrated project in Himalaya for identification of problems, study of causes and suggesting remedial measures is warranted. The proposed study will analyse the environment of Himalaya using a combination of remote sensing, GIS and field techniques.

(vi) Analysis of canopy biochemistry of forest ecosystems

Several studies carried out in the past (mainly at NASA) have indicated that absorption spectra of leaf material contain absorption properties attributable to other biochemical constituents. It is shown that the biochemical constituents of foliage and other organic samples can be accurately estimated from their reflectance spectra in near and shortwave infrared regions.

The basis for these results is the occurrence of spectral absorption features associated with organic constituents throughout the visible and infrared regions. Organic compounds absorb radiation in the MIR and in the UV region at fundamental stretching and bending vibrations of strong molecular bands between light alarms, e.g., hydrogen bands associated with carbon, oxygen and nitrogen.

(vii) Water management in command areas

Reservoir operation and regulation and irrigation scheduling water yield assessment by studying catchment hydrologic characteristics, water demand estimates by assessment of crop acreage and water use efficiencies for optimum reservoir operation. Large potential exists to operationalize and automate the procedure by applying GIS and DEM techniques.

(viii) Water demand assessment by evaporation and evapotranspiration studies

A comprehensive model development involving remote sensing (visible and thermal infrared region) and meteorological data using energy balance and evaporation/and evapotranspiration equations. It will also assist in drought assessment and calls for indigenous thermal band development. This will involve the following:

- Rainfall estimation, forecast and monitoring
- Snow mapping and snow-melt run-off estimation
- Hydrologic modeling for rainfall run-off studies
- Integrated watershed management for identifying sites for soil and water conservation activities
- Automated drainage extraction and morphometric analysis

(ix) Flood disaster warning system

Assessing flow and water yield upstream and applying DEM techniques, a flood disaster warning system can be developed on a near real-time basis to assess the zones likely to be affected and which need to be evacuated.

(x) Landslide zonation

High-resolution stereo images are excellent in identification of active and old landslides, potential landslide zones, areas of unstable slopes, lineaments, fractures and joints. This is especially useful in the high relief, inaccessible terrains of the middle and inner Himalayas. Periodic monitoring is also possible with satellite-based data. Blocks vulnerable to landslides can be selected in the Himalayas for such studies:

a. Structural and neotectonic studies in the Himalayas: using high resolution stereo images, subtle changes in the landforms, slope morphology, alluvial fans, recurrence of landslides, etc. which may be indicative of neotectonic movements can be identified in the stereo images. Such areas in the Himalayas can be selected for studies using IRS-1C stereo-data.

b. Geological morphology studies in Himalayas: high-resolution stereo images can help in identifying various glacial landforms, and in assessing snow cover thickness and morphology.

(xi) Application of microwave data

SAR images are known to be immensely useful for structural delineation. Often geological-structural features are better observed on SAR images than on optical VNIR images and photographs. But in very rugged terrains such as the Himalayas, structural interpretations are hampered by image distortions due to lay over and foreshortening. Proper software should be developed to remove these distortions using SAR digital data.

Soil moisture is one of the most important factors for increasing the crop yields. An optimum moisture content plays a vital role in crop growth. Soil moisture becomes harmful when it is present in the soil at around saturation, because almost all the pores remain occupied with water and soil air becomes deficient. Owing to anaerobic conditions, plant growth suffers when soil moisture is below wilting point in the soil. Because not enough water is available for evapotranspiration as well as for microbial activities and solute movement, the nutrient supply to the plants is hampered.

Microwave remote sensing can be used for estimating soil moisture in the soil. A suitable wavelength polarization has to be found which can penetrate the soil beyond 10 cm, so that soil moisture can be estimated in the root zone in almost real time and also without disturbing the soil.

(xii) Mineral identification using remote sensing

Different minerals are sensitive to different wavelengths. A proper study with a very narrow wavelength over a broad spectrum can help in identifying the minerals. Knowledge about mineralogical composition and soil is important as it controls nutrient availability, moisture availability, porosity and many other properties related to plant growth. This study can be effectively taken up with the available imaging spectrometry data.

(xiii) Crop discrimination using remote sensing

Remote sensing has been operationalized for average estimation of major crops like wheat and paddy, especially in those regions where these are grown in large areas. Owing to limitations of resolution, crops with smaller areas could not be taken up in the past for acreage estimation. With the availability of high-resolution IRS-IC data, its is proposed to take up minor crops like potato, mustard, gram and millet for acreage estimation. This will help in proper land-use planning and knowing the areal extent and properties. Sensors with very narrow spectral bands can be very useful for such studies and these may be available from spectrometers.

(xiv) Mapping of acid soils using remote sensing

There is a voluminous work on remote sensing applications for mapping salt-affected lands in the country. There are large areas with acidic soils and there has been no attempt to make use of remote sensing technology for mapping such areas. It is proposed that IRS-1C data may be used to map acidic soils. This will help in finding out the potential of such soils and their reclamation requirements could also be known more precisely.

Water Resources

- Irrigation scheduling
- Demand/supply studies for the command areas

Urban

- Preparation of city satellite images such as utility maps and city guide maps
- Digital classification methods for urban growth monitoring
- Cadastral overlays at 1:15000 scale
- Land Information system

Agriculture and horticulture

- Crop identification in kharif season using microwave data
- Crop yield modelling
- Automation packages for the analysis
- Horticulture crop mapping (area extent/health assessment)

Soil and land degradation

- Database on soil profile information
- Land degradation mapping on 1:50,000 scale, e.g., development of models for the prediction of land degradation
- Soil moisture studies using microwave data up to 10 cm depth (preferably with 60 cm depth)

Integrated studies

- Redefining the action plan based on impact assessment and changed scenario
- Updating of thematic maps on larger scale
- Cadastral overlays on satellite data

I

Glacial studies

- Hydro-geomorphological mapping on 1:50,000/1:25,000 scale digital procedures
- Satellite data draped on DEM (DTM)
- Collateral data
- Glacier inventory (digital database)
- Water equivalent
- Glacier water equivalent estimate
- Glacial melt model
- Glacier-climate studies and global warming
- Avalanche studies

Geology

- Digital database
- Integration with geophysical and collateral data
- Mineral prognastics

Coastal studies

- Phytoplanbton mapping
- Potential zones for fishing
- Derivation of biological parameters

Natural hazards applications

- Floods monitoring and control
- Earthquakes
- Landslides

(c) Training and promotion

Availability of trained manpower is very crucial for operationization and effective utilization of remote sensing technology. It is high time that due consideration and proper thoughts are given to assessing the training needs in the field of remote sensing and GIS in the coming years. The courses at RRSSC may be organized mainly as follows:

1. Digital image processing for remote sensing applications   2 weeks

2.	GIS and remote sensing	3 weeks
3.	Special courses (forestry, geology, land use and others)	2 weeks
4.	Digital photogrammetry	2 weeks

(i) Remote sensing extension capsule

The thrust is to be laid on the applications of remote sensing in the next five-year plan. The DOS has reached a mature state in many of the areas. Knowledge of remote sensing applications has to reach various levels, e.g. researchers, students, government agencies, NGOs. The proposal is to develop systems to capture the expertise of studies and operational missions such as IMSD, CAPE, LANDUSE/LANDCOVER and many others.

The multimedia system will be best suited for this to develop audio and video materials for propagation and training of remote sensing applications. The multimedia system will also be used to capture the FLY output using EASI/PACE package. The output of the multimedia system may be planned on CD-ROMs and video cassettes to be played on computer and VCRs.

(d) Software development

(i) Automated training system

Development of an automated software package for training in digital image processing and GIS is proposed. The software will be developed in Windows environment. This will help in training the course participants organized by RRSSCs and other institutions.

(ii) Expert system for soil mapping

An expert system for soil mapping and soil conservation survey is proposed. This would enable the users to generate the maps digitally with the help of a soil expert's knowledge. The expert knowledge will be framed and stored in a data bank. The prototype will be developed on expert system shells which are available commercially.

(iii) Image processing software lab

The packages developed by RRSSCs will be integrated under one umbrella with the standards defined. The different functions generated will be made available in two forms. The first option will be the complete software which can be executed; secondly, functions can be called within a programme to meet the future needs. This lab will be UNIX-based and will have interoperability on different platforms.

This will include advanced classification, filtering pattern matching and others; algorithms will be developed in the ninth five-year plan.

(iv) Map, scanning, conversion and management system

For problems of map maintenance and conversion to make compatible for image processing and GIS, an integrated tool to enable the map conversion using scanner, archival, distribution and integration to the database development is proposed. The database will be archived on optical disks which is an effective solution for long-term planning. A management system is to be built in for proper utilization of data to minimize the efforts required for regeneration of data.

(v) Radar interferometry for topographic mapping

Successive orbits of SAR data can be used to derive interferogram. The correlation properties of radar echoes can also be used for inference on changes in surface behaviour.

(vi) Photogrammetry

Automatic slope generation can be done using softcopy photogrammetry. This can be a part of an integrated production system comprising image processing, GIS and expert systems and generation of ortho-images.

(vii) Universal data exchange format

The development of Universal Exchange Format for Digital Cartographic Database should be also similar lines as the Digital Chart of the World. This format can be used as a basis for data archival, retrieve and analysis under the NRIS concept.

(viii) Advanced image processing techniques

With the availability of high resolution (spectral and spatial) data, new techniques have to be developed for processing the data. The following are the areas where RRSSCs can contribute:

- Knowledge-based classification for multi-frequency SAR imageries
- Shape from shading algorithms with ground calibration
- Modified efficient MLE for classifying imaging spectrometer data
- Genetic algorithms as an extention to artificial neural networks
- Data compression techniques for high resolution multispectral data

I

4.2.5 Indonesia

National Institute of Aeronautics and Space

The National Institute of Aeronautics and Space (LAPAN) is a government organization mandated to coordinate and integrate all matters related to space activities in Indonesia, undertake research and development work in space technology applications and manage the remote sensing data archives.

Professional Personnel

Primary field of work	Number of professional staff with practical experience in			Number of staff with the following qualifications			Number of staff with intensive training (>2 months) during last 5 years in		
	RS	GIS	GPS	Ph.D.	MSc	BSc	RS	GIS	GPS
Agriculture/forestry/rangeland management	10	4			3	11			
Ecology/biology/biodiversity									
Geology/geomorphology/soil sciences	5	1			1	5			
Geography	4	1		1		3			
Meteorology/climatology	5				2	3			
Oceanography and fisheries	7				1	6			
Environmental monitoring	10				2	8			
Urban/regional planning	6					6			
Hydrology/water management	2					2			
Rangeland management									
Photogrammetry/cartography/ digital mapping	4		4			4			
Sensor development									
Hardware/software development	10			3	2	6			
Data management/information systems services	10				2	8			
Operation and maintenance	5				2	3			

Total number of professional staff: 83   Total number of administrative staff: 50

Major Projects/Activities (1992-1996)

Title	Geographical extent/surface area in km²	Objectives	Generated products	Staff time	Names of other agencies involved in project/activity
Remote sensing ground station upgrading		To upgrade the capability of LAPAN remote sensing ground receiving station to receive, record and process Landsat, ERS-1 data	Landsat, SPOT, ERS-1, NOAA data and value added products	15 person-years	BPPT, Forest Ministry, Agriculture Ministry, Mining and Energy Ministry, Transmigration Ministry

Title	Geographical extent/surface area in km²	Objectives	Generated products	Staff time	Names of other agencies involved in project/activity
JERS-1 ground station upgrading		To upgrade the capability of LAPAN remote sensing ground receiving station to receive, record and process JERS-1 data	JERS-1 data	8 person-years	NASDA

Major Current/Ongoing Projects/Activities

Title	Geographical extent/surface area in km²	Objectives	Generated products	Staff time	Names of other agencies involved in project/activity
Development and operation of LAPAN remote sensing ground receiving station		To provide and distribute remotely sensed data for activities in various sectors in Indonesia	Satellite data	25 person-years	
Land resources and environment assessment		To provide information on the current condition of natural resources and environment	Value added data	50 person-years	Agriculture Ministry, State Ministry of Environment, etc.
Earth observation centre development		To provide information for monitoring physical processes and phenomena in the atmosphere, in the sea and on land		50 person-years	Agriculture Ministry, State Ministry of Environment

Seminars/Conferences/Workshops Organized (1992-1996)

Type (seminar, conference, workshop, etc.)	Theme/title	Co-sponsorship (if any)	Number of participants	Title of output (publication and/or other)
Seminar	Landsat seminar	MDA, Canada and EOSAT	200	Seminar proceedings
Seminar	Policy/decision maker seminar	MDA, EOSAT and SPOT Image	200	Seminar proceedings
Training workshop	Policy/decision maker seminar, Vancouver, Canada	MDA	15	
Workshop	Radar applications seminar	ESA	50	

Documents Published since 1990

Title	Year of publication	Language of publication	Keywords	Frequency
LAPAN Remote Sensing Research and Development Proceedings	1990-1995	Indonesian	Research and development activities and results	Every year
Berita Bank Data (Remote Sensing Data Bank News)	1994	Indonesian	LAPAN remote sensing activities and data catalogue	Every quarter

Training Activities (1992-1996)

Course title and description	Training undergone by organization staff		
	Given by (name of institution/agency)	Duration/ frequency	Number of staff taking course
Remote sensing ground station operation and maintenance	MDA, Canada	1 year	10
Remote sensing data application	MDA	1 month	5
Quality control assessment	ACRES, Australia	1 month	5
JERS/ADEOS training course	NASDA, Japan	1 week	2
Tropical resources management	ESCAP, NASDA	1 week	2
Application of ERS-1 SAR for coastal zone monitoring	ESA	1 week	4
Radar processing system	DRL, Germany	6 months	2

Course title and description	Training provided by organization to external users		
	Name of participating agencies/organizations	Duration/ frequency	Number of participants
Remote sensing data applications to oceanography	IPB, ITB, UNHAS, GAMA	2 months	20
Radar applications training	Forest Ministry, BPPT, BPPIT, IPB	1 month	25
Remote sensing data applications (introduction)	Forest Ministry, BPN, Transmigration Ministry	1 month	25
SAR and Landsat TM mapping	Forest and farming ministries and private companies	2 weeks	28

Databases/Archives Developed and/or Maintained and Available to Outside Users

Type (digital or analogue) and storage medium	Database contents (type, location, source data, scale and other useful information)	Geographical extent/surface area in km^2	Year of production/ review	Internet address (URL)	Main users
Digital/ hard copy	Landsat TM data archive of Indonesia, Malaysia and parts of Australia and the Philippines	15 million	1993 onwards		Government and private agencies
Digital/ hard copy	SPOT data archive of Indonesia, Malaysia and parts of Australia and the Philippines	15 million	1993 onwards		Government and private agencies
Digital/ hard copy	ERS-1 data archive of Indonesia, Malaysia and parts of Australia and the Philippines	15 million	1993 onwards		Government and private agencies
Digital/ hard copy	JERS-1 data archive of Indonesia, Malaysia and parts of Australia and the Philippines	15 million	1995 onwards		Government and private agencies
Digital/ hard copy	NOAA data archive of Indonesia, Malaysia and parts of Australia and the Philippines	15 million	1981 onwards		Government and private agencies
Digital/ hard copy	GMS data archive of Indonesia, Malaysia and parts of Australia and the Philippines	15 million	1992 onwards		Government and private agencies

Facilities and Activities in other Government and Private Institutions

Organization	Hardware	Software	Professional personnel
National Coordinating Agency for Survey and Mapping	Four workstations and 25 PCs. Peripherals include digitizer, plotter, image capture GPS and other survey equipment.	Erdas, ILWIS, VI²STA, microBRIAN, Earthview, Idrisi, Arc Info, Genasys and Autocad	157
Agency for Assessment and Application of Technology	Hardware installations include workstations, PCs, digitizers and plotters.	VI²STA, Genasys, Arc Info, ERDAS, Idrisi, Multiscope, ILWIS	21
Department of Forestry	Hardware installations include a micro-Vax, PCs, plotters, printers, visual image processing systems, GPS receivers and a photolab.	VI²STA, ILWIS, SPANS, Arc Info	31
Department of Public Works	Hardware installations include a workstation, PCs, digitizer, colour hard copiers and printers.	ERDAS, Arc Info	21
Department of Mines and Energy	Hardware installations include a micro-Vax, PCs, CCD camera system, film writer, colour image scanner and a plotter.	LIPS, Spider, Mapinfo	29
Department of Agriculture	Hardware installations include PCs and a Vax minicomputer.	ArcView, NOAA/NVIDIS and SSTDIS, and Arc Info	20
Indonesia Disaster Management Centre	Hardware installations include PCs, digitizer and printer.	SPANS	2
Department of Transmigration	Hardware installations include PCs, a digitizer, a plotter, a printer and GPS receivers.	Arc Info, Erdas, ILWIS	9
National Land Agency	Hardware installations include a workstation, PCs, digitizers, scanners, a plotter and laser, inkjet and postscript printers.	ERDAS Imagine, ILWIS, Arc Info	18
Geotechnology Research and Development Centre	Hardware installations include a workstation, PCs, digitizers, a scanner and laser, inkjet and postscript printers.	ERDAS Imagine, ILWIS, Arc Info	13
Oceanological Research and Development Centre	Hardware installations include a workstation and PCs.	MicroBRIAN, VI²STA, Arc Info	12

Local governments of the following provinces are each equipped with PCs running Arc Info and ERDAS, digitizers and a printer: Aceh, Bali, Bengkulu, Irian Jaya, Jambi, West Java, Central Java, East Java, Lampung, Riau, South Sulawesi, North Sulawesi, West Sumatra, South Sumatra, North Sumatra, Yogyakarta, West Kalimantan, East Kalimantan. Each provincial government has four professional personnel trained in remote sensing, GIS and satellite-based positioning.

The following universities are each equipped with workstations and PCs running ILWIS, Idrisi and Arc Info, a plotter, a printer and a digitizer: Bogor Agriculture University, Bandung Technology Institute, National Technology Institute, University of Diponegoro, Gadjah Mada University, University of Indonesia, Trisakti University, Udayana University.

The Armed Forces are equipped with a workstation and PCs running OCAPI, ILWIS, ERDAS, Arc Info and Mapinfo. They have 35 professional staff trained in remote sensing, GIS and satellite-based positioning.

There are at least 24 private agencies taking part in remote sensing and GIS application activities.

4.2.6 Islamic Republic of Iran

Iranian Remote Sensing Centre

The Iranian Remote Sensing Centre (IRSC), as the main coordinator for remote sensing, GIS and satellite-based positioning activities in the country has interaction with over 300 government and private sector agencies and educational institutions and provides them with various services. IRSC has played a leading role developing remote sensing and GIS technologies in the country. It has thus been appointed to coordinate all tasks at national, regional and international levels, including tranferring technology to the country. Its organization chart is shown in figure 4.2.

Professional Personnel

Primary field of work	Number of professional staff with practical experience in			Number of staff with the following qualifications			Number of staff with intensive training (>2 months) during last 5 years in		
	RS	GIS	GPS	Ph.D.	MSc	BSc	RS	GIS	GPS
Agriculture/forestry/rangeland management	6	2	2		6	2	2	2	
Ecology/biology/biodiversity									
Geology/geomorphology/ soil sciences	3	3			3	2	1	1	
Geography	4	3	1		5	1	1	2	
Meteorology/climatology	1					1			
Oceanography and fisheries	1			1		1			
Environmental monitoring									
Urban/regional planning									
Hydrology/water management									
Rangeland management									
Photogrammetry/cartography/ digital mapping	4	1			2	1	1		
Sensor development	2				1				
Hardware/software development	3	3				4	1	1	
Data management/information systems services		2			2				
Physics	4					2			
Photography	5	1				2			

Total number of professional staff: 51   Total number of administrative staff: 26

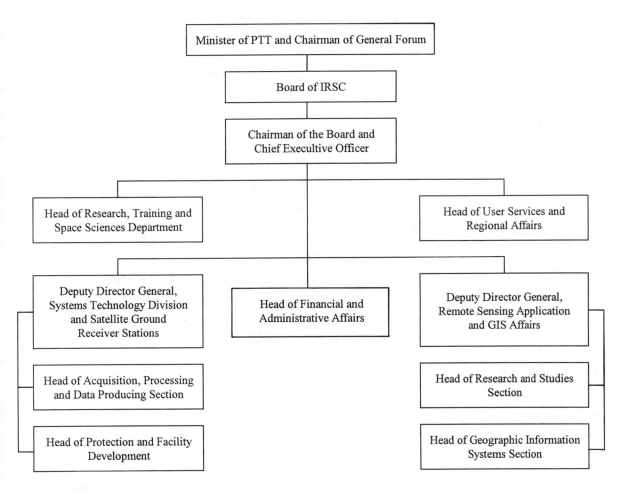

Figure 4.2 Organization chart of the Iranian Remote Sensing Centre

Major Projects/Activities (1992-1996)

Title	Geographical extent/surface area in km^2	Objectives	Generated products	Staff time	Names of other agencies involved in project/activity
ESCAN 90	1,648,000	Land-use mapping	Maps – 350 sheets at 1:100,000 scale	80 person-years	Ministry of Housing and Urban Planning
National atlas	1,600,000	Land resources mapping	Maps at 1:100,000 scale	3 person-years	National Cartographic Centre
ESCAN 90	1,648,000	Mapping of geological faults in the country	Maps – 350 sheets at 1:100,000 scale	20 person-years	Ministry of Housing and Urban Planning
Meshad area study	35,000	Land-use and landform mapping	Maps at 1:100,000 scale	5 person-years	
Rayen project	35,000	Land evaluation	Maps at 1:100,000 scale	5 person-years	

Islamic Republic of Iran

Major Current/Ongoing Projects/Activities

Title	Geographical extent/surface area in km²	Objectives	Generated products	Staff time	Names of other agencies involved in project/activity
Developing GIS of Sistan-Baloochestan Province	134,000	To develop database and GIS of the area	Digital data, maps, data bank	8 person-years	
Fisheries study		To obtain sea surface temperature (SST) and waterfront maps	Maps, digital data	1 person-year	
Change detection study of Sarakhs	16,000	To monitor land-use changes and urban development	Maps	5 person-years	
1 km land-cover database of Asia	Whole country	To develop 1 km land-cover database of Asia. To provide data of Islamic Republic of Iran for incorporation in the database	Maps, digital data	2 person-years	
Land-cover assessment and monitoring of Iran using NOAA AVHRR data	Whole country	To monitor land-cover changes and identify "hot spots"	Maps, digital data	3 person-years	
Research in landslides using remote sensing and GIS	2,000	To evaluate capability of remote sensing and GIS in landslide detection	Maps, digital data	3 person-years	
Study of snow melting	150,000	To map snowcover using NOAA AVHRR data	Maps	1 person-year	Ministry of Power
Development of database and GIS for environmental protection		To establish database in GIS to assist in environmental protection and sustainable development	Digital data	5 person-years	Environment Protection Agency, Sustainable Development Committee and private sector agencies

Seminars/Conferences/Workshops Organized (1992-1996)

Type (seminar, conference, workshop, etc.)	Theme/title	Co-sponsorship (if any)	Number of participants	Title of output (publication and/or other)
International conference	Space science applications	ESCAP	>300	Proceedings and resolution
International conference	Forteenth Asian Conference on Remote Sensing	Asian Association of Remote Sensing	>600	Proceedings
National seminar	Remote sensing applications		220	Final report
National seminar	GIS and GPS			
National workshop	Remote sensing activities	National Fair Organization	250	
National workshop	GIS			
Workshop	Remote sensing and GIS applications in sustainable development			

Type (seminar, conference, workshop, etc.)	Theme/title	Co-sponsorship (if any)	Number of participants	Title of output (publication and/or other)
Workshop	Role of remote sensing in earth resources mapping			
Workshop	On ECO cooperation		>50	
Workshop	Communication ministries of Islamic countries		132	

Training Activities (1992-1996)

Course title and description	Training undergone by organization staff		
	Given by (name of institution/agency)	Duration/ frequency	Number of staff taking course
Training course on the integrated use of remote sensing and GIS for land-use mapping	Yogyakarta, Indonesia	2 months	2
Training course on remote sensing and GIS	Wuhan Technical University of Surveying and Mapping	1 year	3
Training course on pre-processing of NOAA AVHRR data	AIT, Thailand	2 weeks	1
Training course on remote sensing and GIS applications to tropical ecosystem management	ESCAP/NASDA at (a) Bali, Indonesia and (b) Subic, Philippines	1 week each	3

Course title and description	Training provided by organization for external users		
	Name of participating agencies/organizations	Duration/ frequency	Number of participants
Fundamental and advanced course of remote sensing	Forest and Rangeland Research Institute	4 weeks	
Fundamental and advanced course of remote sensing	Ministry of Agriculture	4 weeks	
Fundamental and advanced course of remote sensing	Ministry of Power	4 weeks	
Fundamental and advanced course of remote sensing	Fishery Research Organization	4 weeks	
Fundamental and advanced course of remote sensing	Sistan-Baloochestan province	4 weeks	
Fundamentals of remote sensing	Disaster Committee	2 weeks	
Fundamentals of remote sensing	Geology Surveying Organization	2 weeks	
Fundamentals of remote sensing	Soil Research Institute	2 weeks	
Fundamentals of remote sensing	Shiraz Ostandari	2 weeks	
Fundamentals of remote sensing	Ministry of Jihad	2 weeks	
Fundamentals of remote sensing	Watershed Protection Bureau	2 weeks	
Fundamentals of remote sensing	Water Research Organization	2 weeks	
Fundamentals of remote sensing	Ministry of Interior	2 weeks	
Fundamentals of remote sensing	Environment Protection Organization	2 weeks	
Fundamentals of remote sensing	Shiraz Agricultural Organization	2 weeks	
Fundamentals of remote sensing	Sand Fixation Bureau	2 weeks	

I

Databases/Archives Developed and/or Maintained and Available to Outside Users

Type (digital or analogue) and storage medium	Database contents (type, location, source data, scale and other useful information)	Geographical extent/surface area in km^2	Year of production/ review	Internet address (URL)	Main users
Analogue	Landsat 4, 5 TM data; negative and positive film, 1:1,000,000, 1:500,000 and 1:3,333,000 scale	Whole country	1973-1974, 1984-1985, 1987-1992		Iranian users
Analogue	Landsat 4, 5 TM data; B and W and false colour paper prints, 1:1,000,000, 1:500,000, 1:250,000 and 1:100,000 scale	Whole country	1973-1974, 1984-1985, 1987-1992		Iranian users
Digital	Landsat 4, 5 TM data; CCT, CD-ROM, partly on floppy diskettes	Whole country	1973-1974, 1984-1985, 1987-1992		Iranian users
Digital	SPOT XS and Pan data. CCT	Some parts of central and south	1992-1995, 1997		Iranian users
Analogue	SPOT XS and Pan data; film/prints at 1:50,000 and 1:25,000 scale				Iranian users

Documents Published since 1990

Title	Year of publication	Language of publication	Keywords	Frequency
Many research papers, reports, etc. by the Iranian Remote Sensing Centre	From 1973	Persian/ English		Every quarter

4.2.7 Japan

National Space Development Agency of Japan

The National Space Development Agency (NASDA) of Japan is a government organization mandated to contribute to space exploration and the utilization of the practical applications of space technology for peaceful purposes. Based on the Space Development Basic Plan formulated by the Prime Minister, the involvement of NASDA in space activities includes development of satellites (including space experiments and space station), launch vehicles, satellite tracking technology; and of methods, buildings and facilities required for these activities. Its organization chart is shown in figure 4.3.

Professional Personnel

Primary field of work	Number of professional staff with practical experience in			Number of staff with the following qualifications			Number of staff with intensive training (>2 months) during last 5 years in		
	RS	GIS	GPS	Ph.D.	MSc	BSc	RS	GIS	GPS
Agriculture/forestry/rangeland management									
Ecology/biology/biodiversity									
Geology/geomorphology/ soil sciences									
Geography									
Meteorology/climatology	4			4					
Oceanography and fisheries	3			3					
Environmental monitoring									
Urban/regional planning									
Hydrology/water management									
Rangeland management									
Photogrammetry/cartography/ digital mapping									
Sensor development	20				5	15			
Hardware/software development	28			4	4	20			
Data management/information systems services	10				1	9			
Radiophysics and geomagnetism	3			3					
Earth sciences	5			5					

Total number of professional staff: 23   Total number of administrative staff: 10

Japan

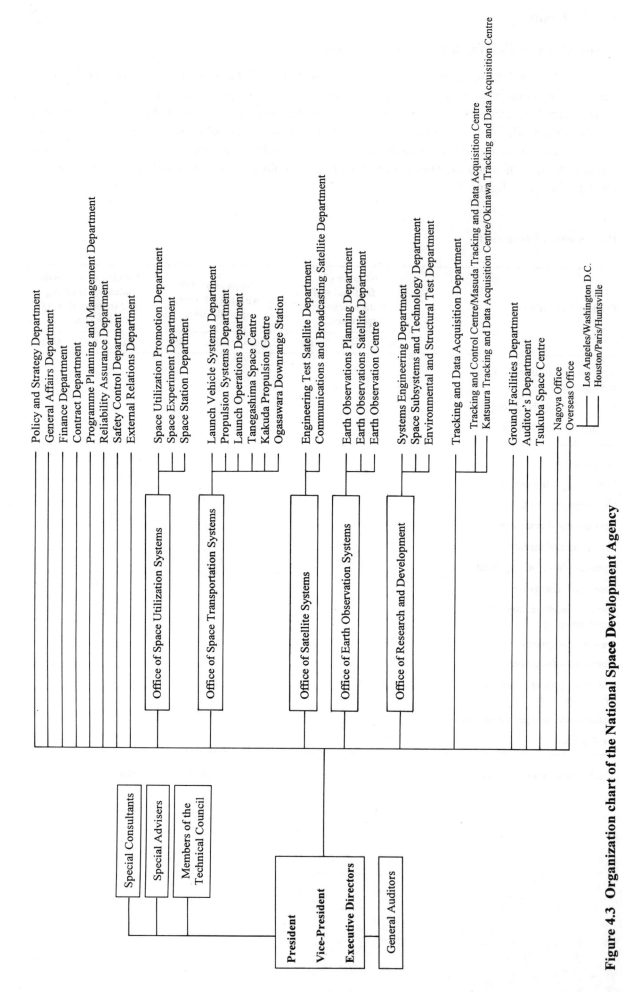

Figure 4.3 Organization chart of the National Space Development Agency

132

Major Projects/Activities (1992-1996)

Title	Geographical extent/surface area in km²	Objectives	Generated products	Staff time	Names of other agencies involved in project/activity
Joint ESCAP-NASDA ADEOS research programme		To demonstrate the potential of ADEOS for environmental and resource monitoring through research studies in various developing ESCAP countries			ESCAP and concerned agencies in 15 developing countries of the region
ADEOS series		To further enhance and demonstrate potential of ADEOS for earth observations through ADEOS calibration/validation, algorithm development and data set generation			
NASA-NASDA TRMM		To conduct scientific investigations relating to tropical rainfall measurements through the TRMM satellite			NASA
GEWEX GAME programme		To investigate the role of the Asian monsoon in the global climatic system			WRCP and others
ADEOS/OCTS DTL validation research		To validate direct transmission to local users system for marine production and fisheries environment monitoring			Ministry of Agriculture, Forestry and Fisheries
GRNS project		To conduct research for preparing data sets for hydrology, vegetation, desert and ocean monitoring/study, and to promote international joint research projects			Science and Technology Agency of Japan and others
Earthquake remote sensing frontier research		To use SAR interferometry techniques, seismic data and electromagenetic phenomena observations for earthquake research			Science and Technology Agency of Japan and others
JERS-1 GRFM project		To identify and map global tropical rainforests			NASA/JPL, Joint Research Centre of European Commission

Major Current/Ongoing Projects/Activities

Title	Geographical extent/surface area in km²	Objectives	Generated products	Staff time	Names of other agencies involved in project/activity
ADEOS science programme		To develop algorithms, conduct calibration/ validation studies, generate data sets			
TRMM project research		To develop algorithms, generate data sets, conduct calibration/validation studies, especially for precipitation radar			NASA, CRL
ADEOS-2, AMSR project and GLI project		To develop algorithms for standard products for the AMSR and GLI instruments			

Seminars/Conferences/Workshops Organized (1992-1996)

Type (seminar, conference, workshop, etc.)	Theme/title	Co-sponsorship (if any)	Number of participants	Title of output (publication and/or other)
Symposium	ADEOS symposium		98	
Workshop	Workshop on advancement of research for chlorophyll observations by ADEOS/OCTS		46	
Conference	GEWEX Asian monsoon experiment international science panel conference		20	
Symposium	ADEOS-2 symposium		80	
Workshop	ADEOS-2 AMSR/GLI workshop		110	

Training Activities (1992-1996)

Course title and description	Training undergone by organization staff		
	Given by (name of institution/agency)	Duration/ frequency	Number of staff taking course
None			2

Course title and description	Training provided for external users		
	Name of participating agencies/organizations	Duration/ frequency	Number of participants
Tropical ecosystem seminar (at Kuching, Malaysia)	From 20 countries	7 days	113
Indonesian remote sensing engineers training (at EOC)	LAPAN, Indonesia	10 days	3
Thai remote sensing engineers training (at AIT)	NRCT, Thailand	12 days	10

Course title and description	Training provided for external users		
	Name of participating agencies/organizations	Duration/ frequency	Number of participants
South-East Asia engineers training No. 1 (at EOC)	JERS-1 operators in China remote sensing satellite ground receiving station	5 days	10
Tropical ecosystem seminar (at Bali, Indonesia)	ESCAP, NASDA, UNCRD, LAPAN invitees and participants from 21 countries	6 days	119
South-East Asia engineers training No. 2	Chinese Science and Technology Committee engineers	10 days	3
Thai remote sensing engineers training (at NRCT Headquarters)	NRCT	12 days	16
South-East Asia engineers training No. 3 (at EOC) study tour for Thai engineers	NRCT, several Ministries	5 days	10
Thai station operation engineers training	NRCT and RESTEC operators	5 days	5
GIS training at AIT	AIT graduate school students	12 days	20
Thai remote sensing engineers (at NRCT Headquarters)	NRCT, several ministries, Kasetsart and Khonkaen universities	2 weeks	15
Operation engineers training for Thai and Indonesian stations	NRCT and LAPAN	2 days	4
Operation engineers training for Thai station	NRCT and RESTEC	5 days	7

Databases/Archives Developed and/or Maintained and Available to Outside Users

Type (digital or analogue) and storage medium	Database contents (type, location, source data, scale and other useful information)	Geographical extent/surface area in km^2	Year of production/ review	Internet address (URL)	Main users
Digital, CD-ROM	JERS-1 science programme-applications data set 1996		1995	ftp.eorc. nasda. go.jp/	
Digital, CD-ROM	JERS-1/SAR rainforest monitoring, Borneo mosaic data set 1996		1995	ftp.eorc. nasda. go.jp/	
Digital, CD-ROM	Satellite image browse data sets, Japan islands		1996	ftp.eorc. nasda. go.jp/	
Digital, CD-ROM	ADEOS/OCTS first images		1996	ftp.eorc. nasda. go.jp/	
Digital, CD-ROM	ADEOS/AVNIR first images		1996	ftp.eorc. nasda. go.jp/	
Digital, CD-ROM	ADEOS AVNIR/OCTS pre-launch sample data		1996	ftp.eorc. nasda. go.jp/	

Type (digital or analogue) and storage medium	Database contents (type, location, source data, scale and other useful information)	Geographical extent/surface area in km^2	Year of production/ review	Internet address (URL)	Main users
Digital, CD-ROM	ADEOS AVNIR/OCTS image data set 1996		1996	ftp.eorc. nasda. go.jp/	
Digital, CD-ROM	Global research network standardized sample data sets		1996	ftp.eorc. nasda. go.jp/	
Digital, CD-ROM	Satellite image data sets (Earth Observations Research Centre)		1995	ftp.eorc. nasda. go.jp/	
Digital, HDDT	Satellite Earth observation data. Path/row number. Processing level data quality. Cloud cover, etc.		1995	ftp.eorc. nasda. go.jp/	

Documents Published since 1990

Title	Year of publication	Language of publication	Keywords	Frequency
JERS-1 data users Handbook	1994	English	Data users manual	
JERS-1 Operation Interface Specification	1993	English	Ground station interface	
ADEOS AVNIR Data Product Specification	1996	English	Product specification	
ADEOS OCTS Data Product Specification	1996	English	Product specification	
Foreign Ground Stations		English	Satellite to ground station interface	
ADEOS Operation Interface Specification	1997	English	Ground station interface	
NASDA Report (http:// www.eorc.nasda.go.jp/)		English		Yearly
EOC News (http://www.eorc.nasda.go.jp/)		English		Monthly

Ministry of International Trade and Industry

The Ministry of International Trade and Industry (MITI), established in 1949, is in charge of sector-specific issues for various industries and cross-sectoral issues such as trade, environment, energy, and others. MITI operates space programmes for the purpose of promoting space utilization and contributing to stable energy supply. It is also in charge of trade control for security and promotional systems (tax, duty, loan) for stimulating space industries. The Space Industry Division of MITI implements the space-related programmes of MITI, supported by experienced scientists and engineers from national laboratories. For remote sensing applications, the Geological Survey of Japan and the Earth Remote Sensing Data Analysis Centre (ERSDAC) play important roles. The organization chart of MITI is shown in figure 4.4.

Professional Personnel

Primary field of work	Number of professional staff with practical experience in			Number of staff with the following qualifications			Number of staff with intensive training (>2 months) during last 5 years in		
	RS	GIS	GPS	Ph.D.	MSc	BSc	RS	GIS	GPS
Agriculture/forestry/rangeland management									
Ecology/biology/biodiversity									
Geology/geomorphology/ soil sciences	10			2	2	6			
Geography	1			1	1				
Meteorology/climatology									
Oceanography and fisheries									
Environmental monitoring									
Urban/regional planning									
Hydrology/water management									
Rangeland management									
Photogrammetry/cartography/ digital mapping									
Sensor development									
Hardware/software development	4			1		3			
Data management/information systems services	6			1	2	3			
Programme management	2					2			

Total number of professional staff: 23   Total number of administrative staff: 10

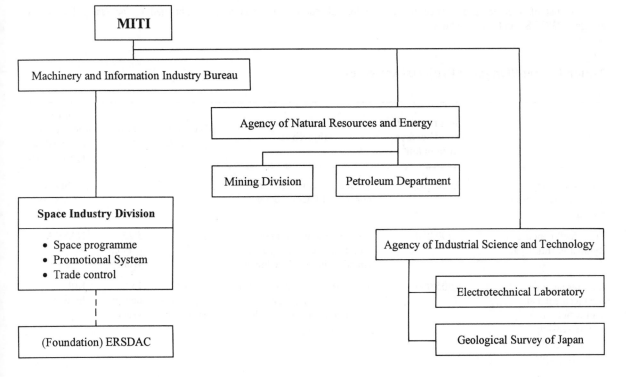

Figure 4.4 Organization chart of MITI with respect to space-related activities

Major Projects/Activities (1992-1996)

Title	Geographical extent/surface area in km^2	Objectives	Generated products	Staff time	Names of other agencies involved in project/activity
Research on geologic remote sensing system	10,000	To specify sensor requirements for geological studies	Report	10 person-years	
Research on advanced geologic remote sensing	200	To define/propose next generation of sensors for geologic use	Report	12 person-years	
Research on techniques for analysing geologic structure from Landsat TM data	10,000	To develop methods and techniques for geologial structural analysis using multi-spectral TM data	Report	8 person-years	Centre for Remote Sensing in Geology (China)
Digitization of geological maps of Japan	Whole country	To make digital geological map of the country at 1:1,000,000 scale	CD-ROM	50 person-years	

Databases/Archives Developed and/or Maintained and Available to Outside Users

Type (digital or analogue) and storage medium	Database contents (type, location, source data, scale and other useful information)	Geographical extent/surface area in km^2	Year of production/ review	Internet address (URL)	Main users
Digital, CD-ROM	Geological map of Japan (1:1,000,000 scale)	Whole of Japan	1995		

A list of documents published by the Geological Survey of Japan may be obtained by accessing http://www.aist.go.jp/GSJ/PSV/Pub/Index-e.htm.

Major Current/Ongoing Projects/Activities

Title	Geographical extent/surface area in km^2	Objectives	Generated products	Staff time	Names of other agencies involved in project/activity
Coral reef mapping in north-west Pacific region	3,000,000	To observe coral reefs in north-west Pacific region	Map		ERSDAC University of Tokyo
Coral reef mapping in Ryukyu Island, Japan	200,000	To map and monitor coral reef distribution in Ryukyu Island using satellite images	Map	5.25 person-years	NASDA
Joint research on enhancement and application of remote sensing Technology with ASEAN countries	1,905,000	To contribute to investigations of non-renewable resources based on analysis of geology and topography	Data of geological structure	15 person-years	LIPI, Indonesia

Title	Geographical extent/surface area in km²	Objectives	Generated products	Staff time	Names of other agencies involved in project/activity
Solid Earth monitoring	Inside the Earth	To study Earth magnetism and electrmagnetic phenomena	Data of Earth's core and mantle	24 person-years	Russia Federation, United States, France
Digitization of geological maps of Japan	Whole country	To prepare digital geological maps of the country at 1:200,000 scale with other geoscientific data			
Research on assessment of geosphere environment in Asia	238,125	To investigate and assess the Asian environment		10 person-years	
ASTER project		To develop the sensor and the analysis method for ASTER/EOS-AH1		20 person-years	ERSDAC, JAROS
AVNIR project		To develop the sensor calibration/validation for ADEOS/AVNIR		7 person-years	NASDA, RESTEC

Earth Remote Sensing Data Analysis Centre

Professional Personnel

Primary field of work	Number of professional staff with practical experience in			Number of staff with the following qualifications			Number of staff with intensive training (>2 months) during last 5 years in		
	RS	GIS	GPS	Ph.D.	MSc	BSc	RS	GIS	GPS
Agriculture/forestry/rangeland management									
Ecology/biology/biodiversity									
Geology/geomorphology/ soil sciences	10			2	2	6			
Geography	1			1	1				
Meteorology/climatology									
Oceanography and fisheries									
Environmental monitoring									
Urban/regional planning									
Hydrology/water management									
Rangeland management									
Photogrammetry/cartography/ digital mapping									
Sensor development									
Hardware/software development	4			1		3			
Data management/information systems services	6			1	2	3			
Programme management	2					2			

Total number of professional staff: 23   Total number of administrative staff: 10

Major Projects/Activities Completed (1992-1996)

Title	Geographical extent/surface area in km²	Objectives	Generated products	Staff time	Names of other agencies involved in project/activity
Application of remote sensing data to oil exploration in southern Sumatra, Indonesia	100,000	To produce geological maps for oil exploration	Base maps oil/gas potential maps	3 person-years	Japex Geoscience Institute, Inc.
Application of SAR data to geological analysis for oil exploration	150,000	To analyse geological structure in Kalimantan, Indonesia	Geological structure maps	3.5 person-years	Japex Geoscience Institute, Inc.
Basic study on SAR applications to tropical rainforest area	80,000	To assess usefulness of SAR for geological mapping in rainforest area	Geological interpretation maps	2 person-years	Japan Energy Development Co. Ltd.
Application of JERS-1 data to logistics and oil exploration in Kazakhstan	120,000	To apply JERS-1 data to logistic and oil exploration	Geological maps Facies map	4.5 person-years	Teikoku Oil Co. Ltd.
Application of JERS-1 data to oil exploration in Bowen Basin, Australia	150,000	To explore oil and gas using JERS-1 data	Geological maps	7.5 person-years	Mitsui Mining and Smelting Co. Ltd. Mitsui Oil Exploration Co. Ltd.

Major Current/Ongoing Projects/Activities

Title	Geographical extent/surface area in km²	Objectives	Generated products	Staff time	Names of other agencies involved in project/activity
Application of JERS-1 data to oil/gas exploration in Junggar Basin, China	150,000	To extract basic geological information for oil/gas exploration	Geological maps Facies map	7.5 person-years	Japex Geoscience Institute, Inc.
Application of radar data to logistics for geophysical exploration in delta area of Bangladesh	80,000	To produce logistics map for seismic exploration	Logistics map	2.5 person-years	Japan Energy Development Co. Ltd.
Application of JERS-1 data to metal exploration in Peru	90,000	To use JERS-1 data for metal exploration	Lineament map	2 person-years	Mitsui Mining and Smelting Co. Ltd.
Application of remote sensing data to metal exploration and to monitoring natural disasters	120,000	To assist in logistics and oil exploration	Landslide map Geological map	5.5 person-years	Dowa Mining Co. Ltd.

Seminars/Conferences/Workshops Organized (1992-1996)

Type (seminar, conference, workshop, etc.)	Theme/title	Co-sponsorship (if any)	Number of participants	Title of output (publication and/or other)
Symposium	International symposium on remote sensing: JERS-1 – towards the future of remote sensing		300	Proceedings
Symposium	International symposium on remote sensing: springboard for the twenty-first century	RESTEC	700	Proceedings
Symposium	International symposium on remote sensing: industrial applications of satellite data and Industrial cooperation		470	Proceedings

Documents Published since 1990

Title	Year of publication	Language of publication	Keywords	Frequency
An Introduction to Resources Remote Sensing	1995	Japanese	History, basic theory, platforms, sensors, image processing, examples of practical applications, future prospects	
Earth Observation Systems from Space	1990	Japanese	Basic theory, rockets and satellites, optical sensors, microwave sensors, data transmission/receiving systems, JERS-1, future prospects	
Data Processing and Analysis of Satellite Imagery (1)	1989	Japanese	Its role and uniqueness, fundamentals, registration, radiometric and geometric corrections, map projection, colour images	
Data Processing and Analysis of Satellite Imagery (2)	1990	Japanese	Spectral analysis, lineament, spectroscopy/DEM, fusion, GIS, hardware and software, neural network, fuzzy, fractakl theory	
Synthetic Aperture Radar (SAR)	1992	Japanese	History, fundamentals, JERS-1 and others, data processing, applications in earth sciences, interferometry, polarimetry	
Case Studies in Resources Exploration (1)	1991	Japanese	Application to petroleum exploration (petroleum geology, oil traps, sedomentary basins, geological structure, lithology, microseepage, logistics)	
Case Studies in Resources Exploration (2)	1993	Japanese	Geomorphology, spectral and thermal information, applications to mineral, geothermal, coal and water resources exploration	

J

Title	Year of publication	Language of publication	Keywords	Frequency
(Supplement Volume) Glossary	1989	Japanese		
Glossary (revised edition)	1996	Japanese		
A report on the progress in remote sensing technology	1990-1995	Japanese	Remote sensing, resources, environment	Annual
Research on the advances of satellite remote sensing for industrial applications	1992	Japanese	Uniqueness of industrial applications/benefits and limitations, current status and the means to bridge the gap between technological achievements and the real world in the context of various earth science disciplines	
Research to expedite the applications of satellite remote sensing for environmental issues	1993	Japanese	Review of current status, UNEP, databases, air-sea interaction, international cooperation, integration with GIS	
Research on satellite remote sensing data systems applied to global environment	1993	Japanese	Trends of earth observation sensors in the context of various earth science disciplines, networks/ databases, applications to natural disasters	
ERSDAC News (Newsletter)	From 1990	Japanese	Activities, research projects, publications, various events, etc., in ERSDAC, worldwide trends in technology and other related topics	Quarterly

4.2.8 Malaysia

Malaysia Centre for Remote Sensing

The Malaysia Centre for Remote Sensing (MACRES) is a government organization mandated to develop and operationalize space remote sensing and related technologies for resources and environmental management in the country. Its organizational chart is shown in figure 4.5.

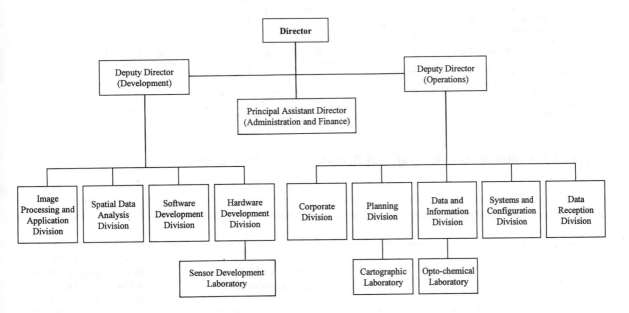

Figure 4.5 Organizational structure of MACRES

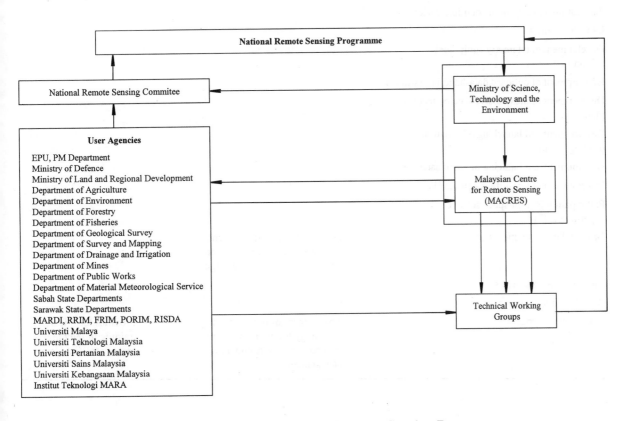

Figure 4.6 Operational framework of the National Remote Sensing Programme

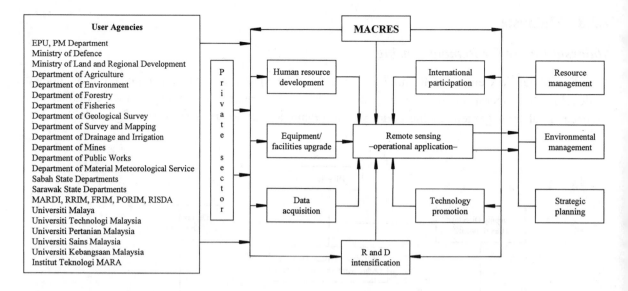

Figure 4.7 Components of the National Remote Sensing Programme

Fields of activity related to space science and technology

Field of activity	Description of major ongoing activities	Description of interaction with developing countries and/or efforts towards addressing their needs
Fabrication and testing of satellites		
Fabrication and testing of satellite launchers		
Fabrication and testing of sounding rockets		
Launch facilities		
Development of telemetry equipment (TT and C)		
Development of onboard data-handling systems		
Development of inexpensive data collection platforms		
Development and launching of equatorial-orbiting satellites		
Development and launcing of small satellites		
Development of scientific satellites (1)		
Participation in joint space science experiments (2)		
Ground data acquisition system	Included in the establishment of the ground receiving system currently at the planning stage.	
Data processing systems		
Data distribution networks		
Data analysis systems	Development of remote sensing technology for applications in resources and environmental management.	Provides technical training in remote sensing to Viet Nam.

Professional personnel involved in space science and technology

Primary field of work	Number of staff with the following qualifications			Number of staff with intensive training (>2 consecutive months) during last 5 years in
	PhD level	Master's level	Bachelor's level	
Fabrication and testing of satellites				(a) Integration of remote sensing and GIS. Attended by 12 officers.
Fabrication and testing of satellite launchers				
Fabrication and testing of sounding rockets				
Launch facilities				
Development of telemetry equipment				(b) Training fo environment and resourcers planning. Attended by 6 officers.
Development of onboard data-handling system				
Development of inexpensive data collection platforms				
Development and launching of equatorial-orbiting satellites				
Development and launching of small satellites				
Development of scientific satellites (1)				
Participation in joint space science experiments (2)				
Data acquisition systems				
Data processing systems			3	
Data distribution networks			2	
Data analysis systems	2		15	

Total number of professional staff: 57 persons

Total number of administrative staff: 14 persons

Documents published since 1990

Title	Year of publication	Language of publication	Keyword	Frequency
1. Nik Nasruddin Mahmood and K.F. Loh (1990). "Economics of implementing the Malaysian Natural Resources Evaluation Programme using remote sensing technologies", *Proceedings*: ESCAP/ADP Regional Conference on the Assessment of the Economics of Remote Sensing Applications to Natural Resources and Environmental Development Projects, Guangzhou, China, 14-18 November 1990.	1990	English		
2. K.F. Loh (1991). "A technical proposal for crop suitability assessment and planning of part of Bantul, Yogjakarta, using remote sensing and geographic information system", Workshop on Land Use Planning, Yogjakarta, Indonesia, 13 May 1991.	1991	English		
3. K.F. Loh, J. Bolhassan, and N.N. Mahmood (1991). "Soil erosion mapping using remote sensing and GIS techniques for land use planners", ASEAN Seminar on Land Use Decision and Policies: Will Tropical Forest Survive Their Impact? Penang, Malaysia, 28-29 October 1991.	1991	English		

Documents *(continued)*

Title	Year of publication	Language of publication	Keyword	Frequency
4. K.F. Loh, Jimat Bolhassan and Nik Nasruddin Mahmood (1992). "Soil erosion mapping using remote sensing and GIS techniques: Upper Klang Valley, Selangor, Malaysia", *Asia-Pacific Remote Sensing Journal*, July 1992.	1992	English		
5. K.F. Loh (1992). "Soil erosion risk assessment using remote sensing and GIS techniques: Middle Klang River Basin, Selangor, Malaysia", ADB-GIS Seminar, Manila, 2-4 December 1993.	1993	English		
6. N.N. Mahmood, K.F. Loh and Salleh Hj. Supian (1993). "SAR and SPOT data for land cover mapping: their complementary and integrated characteristics", Workshop on SAR Data Analysis and Applications, Bangkok, 2-4 March 1993.	1993	English		
7. K.F. Loh, K.Y. Siew and Zuraimi Suleiman (1993). "The acquisition of soil/land cover information using satellite remote sensing techniques", Conference on Soil Resource Management for Agricultural Development, University of Agriculture, Serdang, 5-6 April 1993.	1993	English		
8. N.N. Mahmood and K.F. Loh (1993). "Operationalization of remote sensing application in Malaysia", Seminar on the Application of Remote Sensing in Earth Sciences in the Tropics, National University of Malaysia, 14-15 December 1993.	1993	English		
9. K.F. Loh, Z. Suleiman and I. Saidin (1993). "Soil mapping using remote sensing and GIS technologies", Seminar on the Application of Remote Sensing in Earth Sciences in the Tropics, National University of Malaysia, 14-15 December 1993.	1993	English		
10. N.N. Mahmood, K.F. Loh, Ibrahim Selamat and Z.A. Hassan (1996). "Updating of landcover maps using SAR Data", Seminar on Technology for Updating Maps Using Remote Sensing, Jakarta, 22-24 July 1996.	1996	English		
11. K.F. Loh, Idris Seadin, Adnan Haji Ismail, Ku Mohd. Noh Haji Ku Ramli and Zuraimi Suleiman (1993). "Soil erosion risk assessment of Sintok/Badak water catchment area using remote sensing and GIS technologies", Soil Science Conference of Malaysia, Penang, Malaysia, 19-21 April 1993.	1993	English		
12. K.F. Loh, Idris Saedin, Adnan Hj. Ismail, Ku Mohd. Noh Hj. Ku Ramli and Zuraimi Suleiman (1993). "Soil erosion risk assessment of the Sintok/Badak water catchment area using remote sensing and GIS technologies", Soil Science Conference of Malaysia, Penang, Malaysia, 19-21 April 1993.	1993	English		
13. Nik Nasruddin Mahmood, K.F. Loh and Adnan Hj. Ismail (1993). "Remote sensing imagery for identifying wetlands", National Conference of Agriculture Drainage, Malacca, Malaysia, 9-12 February 1993.	1993	English		
14. K.F. Loh, N.N. Mahmood and Ku Mohd. Noh Haji Ku Ramli (1993). "The integration of remote sensing and GIS technologies for sustainable development", International Space Year Seminar on Remote Sensing for Tropical Ecosystem Management, Kuching, Sarawak, Malaysia, 8-14 September 1993.	1993	English		

Documents *(continued)*

Title	Year of publication	Language of publication	Keyword	Frequency
15. K.F. Loh, N.N. Mahmood and Ku Mohd. Noh Haji Ku Ramli (1993). "Remote sensing applications for water bodies studies for aquaculture", Workshop on Development of Lake and Reservoir for Fish Production in the ASEAN Region, Malacca, Malaysia, 18-22 October 1993.	1993	English		
16. Jasmi Ab. Talib and H.I. Adnan (1994). "GlobeSAR project in Malaysia", Regional GlobeSAR Seminar, Bangkok, 28 November – 2 December 1994.	1994	English		
17. H.I. Adnan, A. Shahruddin and A.T. Jasmi (1994). "Landcover classification using airborne SAR: a preliminary study of Muda-Merbok, Kedah, Peninsular Malaysia", *Proceedings* of Regional GlobeSAR Workshop, Bangkok, 28 November – 2 December 1994.	1994	English		
18. K.F. Loh (1994). "Land-use/cover mapping in the state of Selangor Malaysia", Decision Makers Seminar on Applications of Remote Sensing and Geo-information System, Langkawi, Malaysia, 12-16 December 1994.	1994	English		
19. K.F. Loh (1994). "Development of methodology to monitor tropical forest land cover change", First ADEOS Symposium, Kyoto, Japan, 5-10 December 1994.	1994	English		
20. K.F. Loh (1994). "Integration of space remote sensing and geographic information system for resources management", National Land Information System (NALIS) Symposium, Kuala Lumpur, 14-16 November 1994.	1994	English		
21. K.F. Loh, Ku Mohd. Noh and Laili Nordin (1994). "Complementary nature of SAR and optical data for land cover mapping", Malaysian Decision Makers Seminar on the Potential Application of ERS-1 Microwave Remote Sensing Data in Malaysia, Kuala Lumpur, September 1994.	1994	English		
22. Adnan Hj. Ismail, Jasmi Ab. Talib and Saharuddin Ahmad (1994)", Land cover/land use classification using airborne SAR: a preliminary study of Muda-Merbok, Kedah, Peninsular Malaysia", presented at Regional GlobSAR Workshop, Bangkok, 28 November – 2 December 1994.	1994	English		
23. Regional GlobSAR Workshop Organized by Canada Centre for Remote Sensing; and National Research Council of Thailand, Bangkok, 28 November – 2 December 1994.	1994	English		
24. N.N. Mahmood (1994). "Remote sensing/GIS technology: status and potential in Malaysia", keynote address at the Seminar on Remote Sensing/GIS: Status and Potential, Kuala Lumpur, 22-23 February 1994.	1994	English		
25. K.F. Loh (1994). " Satellite remote sensing technology: its status and use in Malaysia", Current Issues of Science and Technology 1994, Malaysian Scientist Committee, Petaling Jaya, May 1994.	1994	English		

M

Documents *(continued)*

Title	Year of publication	Language of publication	Keyword	Frequency
26. N.N. Mahmood (1994). "Soil erosion risk assessment using remote sensing and GIS technique", paper presented at the Decision Makers Seminar on Application of Remote Sensing an Geo-information System, Langkawi, Malaysia, 12-16 December 1994.	1994	English		
27. N.N. Mahmood (1994). "Environmental Monitoring of the Hydroelectric project area of Bakun, Sarawak, using satellite remote sensing technique", presented at the Decision Makers Seminar, Langkawi. 12-16 December 1994.	1994	English		
28. N.N. Mahmood (1994). "Implementation of national remote sensing programme: a Malaysian model", presented at the National Seminar for Decision Makers on Potential Applications of ERS-1 in Malaysia, September 1994.	1994	English		
29. N.N. Mahmood (1994). "Recent development in space applications is Malaysia", second session of the Asia-Pacific Regional Space Agency Forum, Tokyo, 31 October – 2 November 1994.	1994	English		
30. N.N. Mahmood (1994). "Space applications in Malaysia", Ministerial Conference on Space Application for Development in the ESCAP Region, Beijing, 19-24 September 1994.	1994	English		
31. K.F. Loh (1994). "Remote sensing application in Malaysia", Annual General Meeting of International Hydrological Institute (National Committee, Department of Drainage and Irrigation Malaysia), 12 April 1994.	1994	English		
32. K.F. Loh, Laili Nordin and Ku Mohd. Noh Haji Ku Ramli, Complementary nature of SAR and optical data for land cover mapping: (a) Decision Makers Seminar, Kuala Lumpur, 1 September 1993. (b) ERS-1 National Training Seminar, Kuala Lumpur, 18-30 August 1995. (c) Final Result Seminar, Suranaree University of Technology, Thailand, 19-21 November 1995.	1995	English		
33. K.F. Loh (1995). "Land use/cover mapping of the state of Kedah, Malaysia, using satellite data", International Seminar on Vegetation Monitoring, Chibai University, Japan, 29-30 August 1995.	1995	English		
34. H.T. Ewe, H.T. Chuah, A. Ismail, K.F. Loh and N. Nasruddin (1995), "Paddy crop monitoring using microwave remote sensing techniques", *Geocarta International*, Vol. 10, No. 3 Hong Kong, China.	1995	English		
35. K.F. Loh, L. Nordin, Ku Mohd. Noh (1995). "Complementary nature of ERS-1, SAR and optical data for land cover mapping in Johor, Malaysia", International Seminar on the Integration of Remote Sensing and GIS Technologies, Jakarta, 6-8 June 1995.	1995	English		
36. R.E. Brown, M.G. Wooding, A.J. Batts, K.F. Loh, Ku Mohd. Noh (1995). "Complementary use of ERS/SAR and optical data for land cover mapping in Johore, Malaysia", ERS Application Workshop, London, 6-8 September 1995.	1995	English		

Documents *(continued)*

Title	Year of publication	Language of publication	Keyword	Frequency
37. N.N. Mahmood, K.F. Loh (1995). "Overview of GlobSAR programme in Malaysia", Second Asia Regional GlobSAR Workshop, Beijing, 9-12 October 1995.	1995	English		
38. L. Nordin, K.F. Loh and Ku Mohd. Noh (1995). "Application of simulated radarsat SAR data in land use/cover mapping in Cameron Highlands, Malaysia", Second Asia Regional GlobSAR Workshop, Beijing, 9-12 October 1995.	1995	English		
39. H.T. Ewe, H.T. Chuah, W.Y. Chin, A. Ismail and N.N. Mahmood (1995). "Classification of land use in Kedah, Malaysia, using fractal analysis", Regional GlobSAR Seminar, Beijing, 9-12 October 1995.	1995	English		
40. N.N. Mahmood and K.F. Loh (1996). "Development of space application in Malaysia", Third Asia-Pacific Regional Space Forum, Tokyo, 13-15 March 1996.	1996	English		
41. N.N. Mahmood, K.F. Loh and Ku Mohd. Noh (1996). "SAR backscatter study and land use updating", UN-EC Microwave Remote Sensing Seminar, Manila, 22-26 April 1996.	1996	English		
42. K.F. Loh, Z. Suleiman and Siti Atika (1996). "Generation of the soil map of south-west Selangor using satellite data, Seminar on Advances in Soil, Plant and Foliar Analysis, Sarawak, Malaysia, 15-17 April 1996.	1996	English		
43. *Newsletter* of the Malaysian Centre for Remote Sensing.	June 1991	English		Twice a year
44. Malaysian Remote Sensing Inventory.	1991, 1996	English		Every 5 years

Training activities undergone by organization staff and provided by organization for external users in past five years

Course title and description	Training undertaken by organization staff		
	Given by	Duration/ frequency	Staff taking course
Outside Malaysia			
1. Remote sensing information fellowship, ESA Institute.	European Space Research Institute (ESRIN)	1 January 1990	Mazlan Hashim
2. Three dimension technology from space imagery, AIT, Bangkok.	ESCAP/UNDP Regional Remote Sensing Programme and Asian Institute of Technology.	9-17 April 1990	(a) Marzuki Mohd. Kassim (b) Jimat Bolhassan
3. Seminar on remote sensing application for oceanography and fishery environment analysis, Qing Dao, Beijing.	ESCAP	7-11 May 1990	Mazlan Hashim
4. Workshop on ERS-1/SPOT application: a complementary approach, AIT, Bangkok.	AIT and Natural Resources Programme and Remote Sensing Laboratory.	11-22 June 1990	Loh Kok Fook
5. Workshop on remote sensing application to desertification/ vegetation mapping, Tehran.	ESCAP/UNDP Regional Remote Sensing Programme and the Iranian Remote Sensing Centre.	21-26 August 1990	Laili Nordin

Training (outside Malaysia) *(continued)*

Course title and description	Training undertaken by organization staff		
	Given by	Duration/ frequency	Staff taking course
6. Seminar on remote sensing application for geotecnic mapping and mineral exploration, Tbilisi, Russian Federation.	ESCAP and Russian Federation	26 September – 6 October 1990	Noor Bakri Endut
7. Training programme on remote sensing, Stockholm University, Sweden.	Sweden	13 May – 14 June 1991	Norliza Mohd. Noor
8. Malaysia-China research project, Remote Sensing Institute, University of Peking, Beijing.	ESCAP, Malaysia and China	11-24 August 1991	Loh Kok Fook
9. 43rd Photogrammetric Week 1991, Institute for Photogrammetry, University of Stuttgart, Germany.	University of Stuttgart, Germany	9 September – 14 November 1991	Mazlan Hashim
10. Seminar on the application of space techniques to control natural disaster, Beijing.	ESCAP and China	23-27 September 1991	Mazlan Hashim
11. SPOT course: SPOT system and application, GDTA, Toulouse, France.	GDTA	14-18 October 1991	Noor Bakri Endut
12. 12th Asean Conference on remote sensing, SEAMEO Regional Language Centre, Singapore.	National University of Singapore	29 October – 3 November 1991	Mazlan Hashim
13. Remote sensing training, Sweden.	Sweden	1-15 March 1992	(a) Nik Nasruddin Mahmood (b) Loh Kok Fook (c) Jimat Bolhassan
14. Remote sensing training, Sweden.	Sweden	16-28 March 1992	(a) Idris Saedin (b) Jaafar Abdullah
15. Remote sensing Malaysia-Sweden project training, Sweden.	Sweden	15-30 March 1992	(a) Loh Kok Fook (b) Idris Saedin
16. Application of remote sensing and GIS in environment and natural resources and monitoring, Feldofling, Germany.	Germany	6 April – 5 May 1992	Jimat Bolhassan
17. ERS-1 radar application course, ESRIN, Frascati, Italy.	ESA and ESRIN	22 April – 8 May 1992	Mohd. Salleh Haji Supian
18. Training programme under the "National spatial planning: a complementary tool for natural resources management project, AIT, Bangkok.	AIT	19-26 August 1992	Jasmi Ab. Talib
19. Seminar ERS-1 and SPOT: a complementary tool for natural resources management and human resources requirement in natural resources information technology, AIT, Bangkok.	AIT, NRCT, ESCAP and GDTA	19-28 August 1992	(a) Jasmi Ab. Talib (b) Adnan Haji Ismail
20. Fourth international training course on remote sensing application to geological sciences, Potsdam and Berlin, Germany (1)	UNOSAP, CDG, GFZ and Institute for Geology, Geographic and Geoinformatics	28 September – 16 October 1992	(a) Jasmi Ab. Talib (b) Adnan Haji Ismail
21. Fourth international training course on remote sensing application to geological sciences, Potsdam and Berlin, Germany (2)	UNDP and Germany	29 September – 16 October 1992	Jasmi Ab. Talib
22. Planning meeting of Canada GlobeSAR Mission, Bangkok.	CCRS and IDRC	24-29 May 1993	Jasmi Ab. Talib

Training (outside Malaysia) *(continued)*

Course title and description	Training undertaken by organization staff		
	Given by	Duration/ frequency	Staff taking course
23. 6th ASEAN expert group meeting on remote sensing, Singapore.	ESCAP	12-14 July 1993	Idris Saedin
24. Remote sensing courses and GIS technologies, University of Sussex, United Kingdom.	Ian MacDonald Associates Hunting Technical Services Limited.	19 July – 23 August 1993	Idris Saedin
25. 2nd ERS-1 symposium: space at the service of our environment, Hamburg, Germany.	Germany	11-14 October 1993	Loh Kok Fook
26. International seminar on remote sensing for coastal zone and coral reef applications, AIT, Bangkok.	AIT, GDTA, ESCAP and Natural Resources Planning and Management of Technology.	25 October – 1 November 1993	Adnan Haji Ismail
27. ERS-1 SAR applications for natural hazard monitoring, geology and land use mapping, College of Engineering UP Diliman, Quezon City, Philippines.	ESA	9-19 November 1993	(a) Jasmi Ab. Talib (b) Zuraimi Suleiman
28. Training course in SAR data processing and simulation generation, CCRS, Ottawa, Canada.	MACRES and Canada	18-26 March 1994	Mohd. Salleh Haji Supian
29. Remote sensing application for land use mapping and planning, Gajahmada University, Yogjakarta, Indonesia.	Government of Indonesia, ESCAP and ITC	7 February – 17 April 1994	Loh Kok Fook
30. The integrated use of remote sensing and GIS for land use mapping, Yogjakarta, Indonesia.	ESCAP	22 August – 22 October 1994	Adnan Haji Ismail
31. Third ISY regional remote sensing seminar on tropical ecosystem, Bali, Indonesia.	Germany	23-28 August 1994	Mazlani Muhammad
32. United Nations Workshop on enhancing social, economic and environmental security through space technology, Graz, Austria.	United Nations and European Space Agency	12-15 September 1994	Loh Kok Fook
33. Workshop on microwave remote sensing application, Beijing.	United Nations	14-18 September 1994	Jasmi Ab. Talib
34. EC-ASEAN Regional Training: ERS-1 SAR applications for land use and coastal zone monitoring, Yogjakarta, Indonesia.	European Community and LAPAN-PUSPIC	27 October – 5 November 1994	(a) Ibrahim Selamat (b) Noorhaidah Ariffin (c) Jamilah Ismail
35. International training course on application of remote sensing and GIS in managing tropical forest and conserving natural resources in South-East Asia, Bogor, Indonesia.	JPA and Germany	15 November – 11 December 1994	Mazlani Muhammad
36. Regional seminar on integrated application of remote sensing and GIS for land and water resources management, Bangalore, India.	United Nations	16-19 November 1994	Zuraimi Suleiman

M

Training (outside Malaysia) *(continued)*

Course title and description	Training undertaken by organization staff		
	Given by	Duration/ frequency	Staff taking course
37. Seminar ERS-1 – new and integrated approaches to remote sensing data analysis, Bangkok.	MACRES	20-31 March 1995	(a) Nik Mazlina Nik Mustapha (b) Norizan Abd. Patah (c) Saiful Bahari Abu Bakar (d) Ahmad Radzi Rafi
38. Training course on remote sensing.	Hunting Technical Service, United Kingdom	6 July – 27 September 1995	(a) Muhamed Kamal Azidy Musa (b) Faizu Hassan (c) Noraini Surip (d) Roslinah Samad
39. Tropical ecosystem Seminar, Subic, Philippines.	ESCAP	4-9 September 1995	(a) Darus Ahmad (b) Chin Wee Yoon
40. Symposium on space technology for improving life on earth, Graz, Austria.	ESCAP	11-14 September 1995	Soo Soong Kwan
41. UN/ESA international Training course for Asia and the Pacific countries on applications of ERS data, Frascati, Italy.	ESCAP	13-25 November 1995	(a) Mansor Abd. Rahaman (b) Soo Soong Kwan

In Malaysia

Course title and description	Given by	Duration/ frequency	Staff taking course
1. Application of remote sensing and geographical information systems in Managing tropical rainforest and conserving natural resources in the Asian region.	Food and Agriculture Development Centre (German Foundation for International Development and ASEAN Institute of Forest Management).	11 July – 8 August 1990	Teng Shee Hua
2. Seminar RF engineering and radio propagation.	MIMOS	12-13 March 1991	Norliza Mohd. Noor
3. Fourth ASEAN expert group meeting on remote sensing.	Phuket, Thailand	29-31 August 1991	(a) Nik Nasruddin Mahmood (b) Jimat Bolhassan
4. Satellite communication systems broadcasting – TVRO, networks, Pt-PT techniques and technology.	Australia	8-25 October 1991	(a) Nik Nasruddin Mahmood (b) Mazlan Hashim
5. Seminar on data security management and technical issues Module 1: security and management Module 2: communication and security	Applied Technology Associates Sdn. Bhd., Singapore	13-14 November 1991	(a) Mohd. Salleh Haji Supian (b) Norlia Mohd. Noor
6. Regional seminar in coastal offshore engineering.	University of Technology Malaysia	9-11 December 1991	Mazlan Hashim
7. Pre-symposium events and international symposium on ecology and engineering (ISEE'94).	University of Technology Malaysia	27 October – 3 November 1994	(a) Norizan Abd. Patah (b) Muhamed Kamal Azidy Musa
8. International training course on application of remote sensing and GIS in managing tropical forest and conserving natural resources in South-East Asia.	German Foundation for International Development Food and Agriculture Development Centre.	5 November – 1 December 1994	Norizan Abd. Patah

Training (in Malaysia) *(continued)*

Course title and description	Training undertaken by organization staff		
	Given by	Duration/ frequency	Staff taking course
9. Short course on LIS/GIS.	ITM, Shah Alam	24-26 November 1994	(a) Ku Mohd Noh Ku Ramli (b) Ahmad Kamel Abd. Ghani
10. SPANS GIS seminar.	Mesiniaga Sdn. Bhd.	29 November 1994	(a) Mariamni Halid (b) Siti Atikah Mohamed Hashim (c) Noraini Surip
11. Short course on remote sensing, University of Technology Malaysia, Johor.	University of Technology Malaysia	–	(a) Jefri Mat Saad (b) Zainul Izzah Ahmad (c) Salmah Kasim
12. International of Japanese management: transferring the experiences to Asia, Kuala Lumpur.	MACRES	4 August 1995	Lee Choo Har

Course title and description	Training provided for external users		
	Name of participating agencies/organizations	Duration/ frequency	Number of participants
1. Radar remote sensing workshop.	User agencies	13 August – 1 September 1990	29
2. Workshop on remote sensing.	User agencies	6-25 May 1991	29
3. Workshop on geographic information systems.	User agencies	2-14 September 1991	30
4. Workshop on advanced remote sensing.	User agencies	22 June – 9 July 1992	27
5. Workshop on global positioning system and geographic information system (with GeoResearch Inc., United States).	User agencies	6-13 November 1992	29
6. Workshop on introduction to remote sensing (with University of New South Wales, Australia).	User agencies	22-27 February 1993	16
7. Short course on radar remote sensing (with University of New South Wales, Australia).	User agencies	22-26 June 1993	26
8. Workshop on integration of remote sensing and geographic information (with University of New South Wales, Australia).	User agencies	6-23 December 1993	24
9. Workshop on remote sensing for urban applications.	User agencies	4-13 April 1994	17
10. Workshop on remote sensing for marine applications (with Institute of Remote Sensing, Peking University, Beijing).	User agencies	1-6 August 1994	16
11. Radarsat workshop.	User agencies	21 August – 2 September 1995	30

M

4.2.9 Mongolia

National Remote Sensing Centre

The National Remote Sensing Centre of Mongolia is a government organization mandated to coordinate remote sensing and GIS activities in the country, to develop a natural resources management system, and to receive and process remotely sensed data.

Professional Personnel

Primary field of work	Number of professional staff with practical experience in			Number of staff with the following qualifications			Number of staff with intensive training (>2 months) during last 5 years in		
	RS	GIS	GPS	Ph.D.	MSc	BSc	RS	GIS	GPS
Agriculture/forestry/rangeland management	2	2	7	2	3	6			
Ecology/biology/biodiversity	7	7	7	2	4	15	3	2	1
Geology/geomorphology/ soil sciences	10	10	10	3	4	23	3	3	3
Geography	5	5	5	2	3	10			
Meteorology/climatology	3	3		1	1	4	1		
Oceanography and fisheries	15	15	15	6	16	13	5	5	4
Environmental monitoring	1	3	6	4	5	4	1	1	
Urban/regional planning	1	3	6	4	5	4	1	1	
Hydrology/water management	3	4	5	2	4	6	1		
Rangeland management	1	1			1	1			
Photogrammetry/cartography/ digital mapping	5	5	20	2	5	23	3	1	1
Sensor development									
Hardware/software development	6	6	6	2	2	14	2	2	1
Data management/information systems services	3	3			2	4	3	3	

Total number of professional staff: 206   Total number of administrative staff: 15

Major Projects/Activities Completed (1992-1996)

Title	Geographical extent/surface area in km^2	Objectives	Generated products	Staff time	Names of other agencies involved in project/activity
Protected area database management	National	To develop a database To produce a GIS map	Digital, maps	3 person-years	Hustain Nuruu, Gorkhi Terelj
Ulaanbaatar city's GIS based database	Local	To develop a database To produce a GIS map	Digital, maps	2 person-years	Municipal Office

Major Current/Ongoing Projects/Activities

Title	Geographical extent/surface area in km²	Objectives	Generated products	Staff time	Names of other agencies involved in project/activity
Environmental database of Mongolia	National	To create a database and produce thematic maps	Map, DEM	7 person-years	Geodesy and Cartography Department
Natural resource management	Local	To create a GIS database of four local areas of two provinces	Digital, map	5 person-years	Denmark (DANIDA)

Seminars/Conferences/Workshops Organized (1992-1996)

Type (seminar, conference, workshop, etc.)	Theme/title	Co-sponsorship (if any)	Number of participants	Title of output (publication and/or other)
Workshop	Receiving and processing of satellite data	NASA	15	User's guide
Workshop	Natural disaster monitoring		35	
Seminar	Natural disaster reduction	Russian and Mongolian civil defence organizations	27	

Documents Published since 1990

Title	Year of publication	Language of publication	Keywords	Frequency
Various research papers and reports in journals and proceedings of conferences/ seminars				

Training Activities (1992-1996)

Course title and description	Training undergone by organization staff		
	Given by (name of institution/agency)	Duration/ frequency	Number of staff taking course
Remote sensing and GIS	WTUSM China (arrranged through ESCAP)	1 year	7
Remote sensing applications	IIRC Government of India	1 year	4
GIS applications	ESCAP	2 weeks	2
Remote sensing and GIS applications	ITC, Netherlands	1 year	3

M

Course title and description	Training provided by organization for external users		
	Name of participating agencies/organizations	Duration/ frequency	Number of participants
Grassland and ecosystem management and sustainable development	Many countries	1 week	56
Natural disaster monitoring	Civil Defence Department	1 week	35
Receiving and processing of satellite data	Institute of Meteorology	2 weeks	15

Databases/Archives Developed and/or Maintained and Available to Outside Users

Type (digital or analogue) and storage medium	Database contents (type, location, source data, scale and other useful information)	Geographical extent/surface area in km²	Year of production/ review	Internet address (URL)	Main users
Analogue (photographic)	NOAA HRPT archive	Lat. 35-80°N Long. 80-130°E	1974-1982		Ministry of Nature and Environment, Ministry of Agriculture
Digital (8 mm, 4 mm, 9-track)	NOAA AVHRR archive	Lat. 35-80°N Long. 80-130°E	1988 to date		Ministry of Nature and Environment, Ministry of Agriculture
Digital	Environmental database (soil, forest, landscape, vegetation, ground and surface water, etc.)	Mongolia	1995 onwards		Many users

4.2.10 Myanmar

Department of Meteorology and Hydrology

The Department of Meteorology and Hydrology, a government organization, provides the representative of Myanmar at the World Meteorological Organization and on the National Environment Commission, the National UNESCO Commission, the National Wildlife and Forest Reserve Committee and the National Intergovernmental Ocean Committee.

Professional Personnel

Primary field of work	Number of professional staff with practical experience in			Number of staff with the following qualifications			Number of staff with intensive training (>2 months) during last 5 years in		
	RS	GIS	GPS	Ph.D.	MSc	BSc	RS	GIS	GPS
Agriculture/forestry/rangeland management									
Ecology/biology/biodiversity									
Geology/geomorphology/ soil sciences						2			
Geography						10			
Meteorology/climatology	3				1	1			
Oceanography and fisheries									
Environmental monitoring									
Urban/regional planning									
Hydrology/water management					1				
Rangeland management									
Photogrammetry/cartography/ digital mapping									
Sensor development									
Hardware/software development									
Data management/information systems services									

Total number of professional staff: 3   Total number of administrative staff: 22

Seminars/Conferences/Workshops (1992-1996)

Theme/title	Co-sponsorshop (if any)	Number of participants	Title of output (publication and/or other
National workshop on integrated applications of GIS and remote sensing	ESCAP	16	Advisory services report

Training Activities (1992-1996)

Course title and description	Training undergone by organization staff		
	Given by (name of institution/agency)	Duration/ frequency	Number of staff taking course
ESCAP/China cooperation on in-depth training for remote sensing and GIS applications	Wuhan Technical University of Surveying and Mapping, China	1 year	1
Training course on the integrated use of remote sensing and GIS for land-use mapping	Gadjah Mada University, Indonesia	2 months	2

Databases/Archives Developed and/or Maintained and Available to Outside Users

Type (digital or analogue) and storage medium	Database contents (type, location, source data, scale and other useful information)	Geographical extent/surface area in km^2	Year of production/ review	Internet address (URL)	Main users
Digital	Hydrological database software		1984		

Documents Published since 1990

Title	Year of publication	Language of publication	Keywords	Frequency
Annual Weather Review		English		Yearly
Hydrological Annual		Myanmar		Yearly
Assessment of Rainfall in Myanmar		Myanmar		Every 10 days, monthly, yearly

4.2.11 Nepal

Forest Research and Survey Centre

The Forest Research and Survey Centre is a government organization which utilizes space-related data for forest resources surveying and mapping and monitoring forest cover changes in the country. In addition, the Centre coordinates all activities concerning RESAP, as the national focal point. Its organizational chart is shown in figure 4.8.

Professional Personnel

Primary field of work	Number of professional staff with practical experience in			Number of staff with the following qualifications			Number of staff with intensive training (>2 months) during last 5 years in		
	RS	GIS	GPS	Ph.D.	MSc	BSc	RS	GIS	GPS
Agriculture/forestry/rangeland management	2				2				
Ecology/biology/biodiversity									
Geology/geomorphology/ soil sciences									
Geography									
Meteorology/climatology									
Oceanography and fisheries									
Environmental monitoring									
Urban/regional planning									
Hydrology/water management									
Rangeland management									
Photogrammetry/cartography/ digital mapping	5				4				
Sensor development									
Hardware/software development									
Data management/information systems services									

Total number of professional staff: Total number of administrative staff:

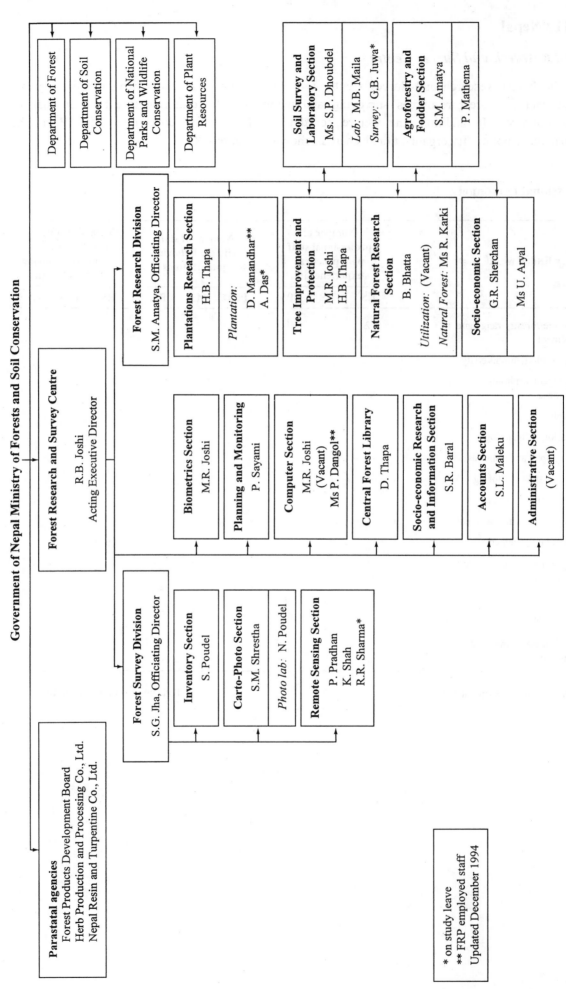

Government of Nepal Ministry of Forests and Soil Conservation

Figure 4.8 Organization chart of the Forest Research and Survey Centre

Major Projects/Activities Completed (1992-1996)

Title	Geographical extent/surface area in km²	Objectives	Generated products	Staff time	Names of other agencies involved in project/activity
Forest resources and deforestation in lowland forests of Nepal	2 administrative districts (35,000 sq. km.)	To find change in forest cover and rate of deforestation in lowland forests of Nepal (20 administrative districts) between 1978 and 1990	Reports, maps	2 person-years	
Preparation of woody vegetation maps of eastern and central development regions of Nepal		To prepare vegetation maps	Reports, maps	0.5 person-year	

Documents Published since 1990

Title	Year of publication	Language of publication	Keywords	Frequency
Various research papers, project reports, status reports, inventories, manuals, etc.	1990-1996	English		

N

4.2.12 Pakistan

Space and Upper Atmosphere Research Commission

The Space and Upper Atmosphere Research Commission (SUPARCO) is an autonomous commission under the federal government. As the national space agency, SUPARCO is responsible for the implementation of the national space programme, which is essentially aimed at furthering research in space science and related fields, enhancing capabilities in space technology and promoting the applications of space science and technology for the socio-economic development of the country. SUPARCO is also the national coordinator for remote sensing activities in the country. Its organizational chart is shown in figure 4.9.

Professional Personnel

Primary field of work	Number of professional staff with practical experience in			Number of staff with the following qualifications			Number of staff with intensive training (>2 months) during last 5 years in		
	RS	GIS	GPS	Ph.D.	MSc	BSc	RS	GIS	GPS
Agriculture/forestry/rangeland management	2	1		1	1				
Ecology/biology/biodiversity									
Geology/geomorphology/soil sciences	2				1	1	1		
Geography	2	1		1	1	1	1		
Meteorology/climatology	4				4				
Oceanography and fisheries									
Environmental monitoring	2			1	1				
Urban/regional planning	2	2			2				
Hydrology/water management	3								
Rangeland management	1	1		1					
Photogrammetry/cartography/ digital mapping									
Sensor development	3				3				
Hardware/software development	2								
Data management/information systems services	5								

Total number of professional staff: 31   Total number of administrative staff: 15

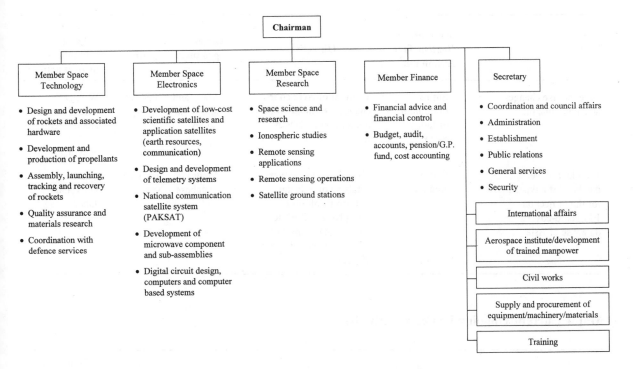

Figure 4.9 Organization chart of the Space and Upper Atmosphere Research Commission

Major Projects/Activities Completed (1992-1996)

Title	Geographical extent/surface area in km²	Objectives	Generated products	Staff time	Names of other agencies involved in project/activity
Possible sites for construction of dams	8 different catchment areas	To identify suitable sites for possible siting of small dams	Maps showing likely sites	2.5 person-years	
Urban land-use/ land-cover classification	Major cities and surrounding areas	To study temporal land-use/ land-cover changes and growth in major cities	Thematic maps	3 person-years continuing	Karachi Development Authority
Wetland mapping	Entire country	To map permanent and temporary wetlands in the country	Thematic maps	2 person-years	
Mapping and monitoring of coastal areas	Coastal belt of Pakistan	To study environmental processes and changes in coastal areas	Maps	2.75 person-years	
Snowcover estimation	Upper Indus basin in northern mountains	To estimate areal extent of snow cover in high mountain ranges (Himalayas/Karakorum/ Hindu Kush) and its correlation with river run-off in Indus basin	Statistical tables, graphs, equations, maps	3.5 person-years	Water and Power Development Authority (WAPDA)
Deforestation in riverine forests	Areas along River Indus in Sindh Province	To monitor and study deforestation and related landcover changes in riverine forests along the Indus in Sindh over a 13-year period	Maps	2 person-years	

Title	Geographical extent/surface area in km^2	Objectives	Generated products	Staff time	Names of other agencies involved in project/activity
River course changes and flood mapping	Entire floodplain of Indus basin	To monitor river course changes, estimate flood inundation and study flood recession	Maps and statistical tables	3 person-years	WAPDA, Irrigation Departments of the provinces
1 km land-cover database of Pakistan for the project on 1 km land-cover database of Asia	Entire country	To develop landcover database of Pakistan at 1 km resolution using NOAA AVHRR HRPT data, as part of 1 km landcover database of Asia	Maps Database	1 person-year	Chiba University, Japan

Major Current/Ongoing Projects/Activities

Title	Geographical extent/surface area in km^2	Objectives	Generated products	Staff time	Names of other agencies involved in project/activity
Urban land-use/ land-cover classification	Major cities of the country	To monitor urban growth, land-use changes and environmental impact	Thematic maps	3 person-years	Concerned civic agencies
Agricultural crop area estimation	Various agricultural areas in Pakistan	To identify and estimate area under different agricultural crops	Thematic maps statistical tables	3 person-years	Pakistan Agricultural Research Council
Waterlogging and salinity monitoring	Affected districts in Punjab and Sindh	To monitor, map and estimate area afflicted by waterlogging and/or salinity	Maps	2.5 person-years	WAPDA, Provincial Agricultural Departments
Mangrove mapping	Indus delta and adjoining coastal areas		Maps	3.6 person-years	Institute of Marine Sciences, Karachi University
Study of turbidity/ sedimentation in dam reservoirs	Hub Dam near Karachi	To study turbidity/ sedimentation and other related aspects in dam reservoirs	Maps	2 person-years	WAPDA
Delineation of ecological zones in Pakistan	Entire country	To classify country ecologically and produce ecological maps	Maps	2 person-years	Soil Survey of Pakistan
Coastal zone monitoring	Coastal areas around Karachi	To investigate various dynamic geomorphic and environmental changes in a fast developing coastal area using ADEOS data	Maps Database for GIS	3 person-years	National Institute of Oceanography, Karachi Development Authority

Note: Planned new activities (not specifically listed above) that the organization aims to undertake in the next 3-5 years include the use of very high resolution optical data from current and future satellites and the use of microwave data from ERS, JERS and Radarsat.

Seminars/Conferences/Workshops Organized (1992-1996)

Type (seminar, conference, workshop, etc.)	Theme/title	Co-sponsorship (if any)	Number of participants	Title of output (publication and/or other)
Meeting	Meeting of directors of national remote sensing centres in the ESCAP region and ninth session of ICC on the RRSP (Islamabad, 1993)	ESCAP/ UNDP, Government of France	40	Proceedings (published by ESCAP)
Workshop	Workshop on microwave/radar remote sensing applications (Karachi, 1993)	CCRS	25	
Seminar	Introductory seminar on remote sensing applications to agriculture (Faisalabad, 1993)	Agriculture University, Faisalabad	120	
Seminar	Seminar on space technology and its significance for Pakistan (Lahore, 1994)	Institute of Engineers, Pakistan	100	Proceedings (published by SUPARCO)
Conference	Second Asia-Pacific conference on multilateral cooperation in space technology and applications (Islamabad, 1995)	Space agencies in China, Republic of Korea, Thailand	110	Proceedings (published by SUPARCO)

Training Activities (1992-1996)

Course title and description	Training undergone by organization staff		
	Given by (name of institution/agency)	Duration/ frequency	Number of staff taking course
Third United Nations international training course on remote sensing applications to geological sciences (Potsdam, Germany, 1991)	United Nations and German scientific and research organizations (DLR, BGR and others)	3 weeks	1
ERS-1 data applications course (Frascati, Italy, 1994)	ESA	3 weeks	2
ERS-1 data applications course (AIT, Thailand, 1996)	ESA, AIT	2 weeks	1
Training workshop on applications of microwave remote sensing to disasters (AIT, Thailand, 1996)	ESA, AIT	2 weeks	2
Master's degree course in remote sensing applications to natural resources management (AIT, Thailand, 1995-1996)	AIT	20 months	1
Postdoctoral United Nations fellowship on remote sensing information systems (ESRIN/ESA, Frascati, Italy, 1995-1996)	ESA	1 year	1

P

Course title and description	Training provided by organization for external users		
	Name of participating agencies/organizations	Duration/ frequency	Number of participants
Fourteenth short training course on remote sensing applications	About 25 government and private sector agencies	2 weeks	54
Fifteenth short training course on remote sensing and GIS applications	About 23 government and private sector agencies	2 weeks	47
Special on-the-job training courses on remote sensing applications	Courses are organized on specific request of national user agencies	2 weeks	10-20
Special eigth-week-long on-the-job training programme on remote sensing and GIS applications	University of Zimbabwe, Harare, Zimbabwe	8 weeks	1

Databases/Archives Developed and/or Maintained and Available to Outside Users

Type (digital or analogue) and storage medium	Database contents (type, location, source data, scale and other useful information)	Geographical extent/surface area in km^2	Year of production/ review	Internet address (URL)	Main users
Digital CCT, HDDT	Database and archive of Landsat MSS/TM and SPOT XS/Pan data of Pakistan and surrounding areas acquired by SUPARCO's satellite ground station (SGS) since 1989. Also some NOAA AVHRR HRPT data	Mainly Pakistan. Also, other areas in zone of acquisition	Data are routinely acquired and archived		National and international user agencies
Analogue prints, transparencies	Landsat and SPOT photographic data of Pakistan and surrounding areas acquired by SGS since 1989. Also, Landsat data of 1972-1988 procured before commissioning of SGS	Pakistan	Data are routinely acquired and archived		National user agencies

Documents Published since 1990

Title	Year of publication	Language of publication	Keywords	Frequency
Space Horizons – Research Journal	1984 onwards	English	Research papers, reports, articles, etc., on various aspects of space science, technology and applications	Quarterly
SUPARCO News (previously SUPARCO Times) – Newsletter on SUPARCO activities	1984 onwards	English	Activities, events, projects, developments, personnel, etc., concerning SUPARCO	Monthly
Space Research in Pakistan, biennial national report submitted to the Committee on Space Research (COSPAR)	Since the 1960s	English	Pakistan's/SUPARCO space programme – organization, establishments, R and D activities, projects, facilities, seminars/conferences, training, international cooperation	Biennial

Title	Year of publication	Language of publication	Keywords	Frequency
Proceedings of the Seminar on Space Technology and its Significance for Pakistan (Lahore, 1994)	1994	English	Space technology, communications, remote sensing for resource/ environmental surveying, satellite meteorology	
Proceedings of the Second Asia-Pacific Conference on Multilateral Cooperation in Space Technology and Applications (Islamabad, 1995)	1996	English	Space technology, its applications, Asia-Pacific regional cooperation, institutional framework, small multi-mission satellite project	
ISNET News, Newsletter covering activities of the Inter-Islamic Network on Space Sciences and Technology coordinated by Pakistan		English	News about activities, events, developments relating to ISNET	Bi-annual
Various research papers, status reports, country reports, opinion papers, etc., published in national and international journals and proceedings of seminars/conferences		English		

P

4.2.13 Philippines

National Mapping and Resource Information Authority

The National Mapping and Resource Information Authority (NAMRIA) is a government organization serving as its central mapping agency. It serves the public and private sectors with regard to mapping, surveying, remote sensing data and natural resources management. Its organizational chart is shown in figure 4.10.

Professional Personnel

Primary field of work	Number of professional staff with practical experience in			Number of staff with the following qualifications			Number of staff with intensive training (>2 months) during last 5 years in		
	RS	GIS	GPS	Ph.D.	MSc	BSc	RS	GIS	GPS
Agriculture/forestry/rangeland management	17	6			3	28	2		
Ecology/biology/biodiversity									
Geology/geomorphology/ soil sciences		1			3	6	1		
Geography		6							
Meteorology/climatology									
Oceanography and fisheries									2
Environmental monitoring	19	11			1		2		2
Urban/regional planning									
Hydrology/water management								2	6
Rangeland management									
Photogrammetry/cartography/ digital mapping		91			2	1	1		13
Sensor development									
Hardware/software development		2						4	
Data management/information systems services	4	6			2	5	12	9	
Geodetic surveying			7						
Business administration, engineering, statistics, economics, architecture				6	99				

Total number of professional staff: 158   Total number of administrative staff: N/A

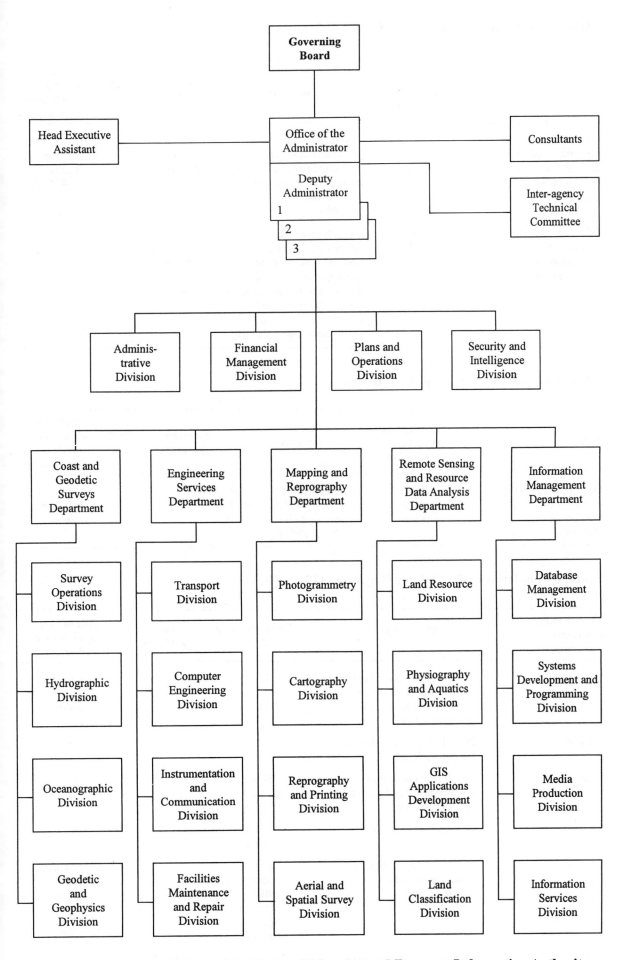

Figure 4.10 Organization chart of the National Mapping and Resource Information Authority

P

Major Projects/Activities Completed (1992-1996)

Title	Geographical extent/surface area in km²	Objectives	Generated products	Staff time	Names of other agencies involved in project/activity
Establishment of a first-order geodetic control network in the Philippines using GPS	300,000 sq. km. (approx.)	To establish and maintain a geodetic control network that would serve as a common reference for the accuracy of all surveys in the Philippines	Geodetic control stations and database		DENR, Land Management Bureau, CERTEZA
Information system project for the management of tropical forests	300,000 sq. km. (approx.)	To study and analyse basic information on the country's land-use, forest distribution, forest resources, deforestation, damage due to forest fires and floods. The results and analyses serve as inputs to forest management planning	Land-use/ forest-type maps and forest registers		

Major Current/Ongoing Projects/Activities

Title	Geographical extent/surface area in km²	Objectives	Generated products	Staff time	Names of other agencies involved in project/activity
Increasing the density of the Geodetic Control Network of the Philippines	300,000 sq. km. (approx.)	To increase density of the existing Geodetic Control Network by having stations at 5 km spacing			Land Management Bureau

Seminars/Conferences/Workshops Organized (1992-1996)

Type (seminar, conference, workshop, etc.)	Theme/title	Co-sponsorship (if any)	Number of participants	Title of output (publication and/or other)
Workshop	EC-ASEAN training workshop on ERS-1 SAR for natural hazard monitoring, geology and landuse (1993)	CEC, ESA	25	Individual country reports
Seminar	EC-ASEAN national seminar for decision makers (1993)	CEC, ESA		
Seminar	Fourth regional remote sensing seminar on tropical ecosystem management (1995)	ESCAP, NASDA, RESTEC		
Seminar	GEOSAR seminar (1995)	INTERA (Canada)	12	
Workshop	United Nations/Philippines workshop on microwave remote sensing applications (1996)	United Nations Office of Outer Space Affairs	80	Individual country reports
Forums	Technical forums on remote sensing, GIS, mapping (yearly)		60-80	

Documents Published since 1990

Title	Year of publication	Language of publication	Keywords	Frequency
Philippines Remote Sensing Newsletter	1990 to present	English	Remote sensing	Annual
INFOMAPPER	1991 to present	English	Surveys, mapping, remote sensing, GIS	Annual
GISLINK	1993 to present	English	GIS	Annual

Databases/Archives Developed and/or Maintained and Available to Outside Users

Type (digital or analogue) and storage medium	Database contents (type, location, source data, scale and other useful information)	Geographical extent/surface area in km²	Year of production/ review	Internet address (URL)	Main users
Digital CCT, Exabyte	Landsat 1, 2, 3 MSS data Landsat 4, 5 MSS data	Philippines	1972-1978 1983-1986		
Digital CCT, Exabyte	SPOT XS data	Philippines	1987-1988		
Digital, analogue diskettes	Land-cover maps (1:250,000) of the Philippines based on SPOT data in Arc Info format	Philippines	1987-1988		
Digital, analogue diskettes	Provincial maps of the Philippines in Arc Info format				

Training Centre for Applied Geodesy and Photogrammetry, University of the Philippines

The Training Centre for Applied Geodesy and Photogrammetry, University of the Philippines is a state educational institution mandated to provide undergraduate education in photogrammetry, geodesy and surveying; provide graduate education in remote sensing and GIS; provide training to private and government staff; and conduct relevant research in the geomatics field.

Professional Personnel

Primary field of work	Number of professional staff with practical experience in			Number of staff with the following qualifications			Number of staff with intensive training (>2 months) during last 5 years in		
	RS	GIS	GPS	Ph.D.	MSc	BSc	RS	GIS	GPS
Agriculture/forestry/rangeland management	2	1			1		2	1	
Ecology/biology/biodiversity	1								
Geology/geomorphology/ soil sciences	2				1				
Geography									

Primary field of work	Number of professional staff with practical experience in			Number of staff with the following qualifications			Number of staff with intensive training (>2 months) during last 5 years in		
	RS	GIS	GPS	Ph.D.	MSc	BSc	RS	GIS	GPS
Meteorology/climatology									
Oceanography and fisheries							1		
Environmental monitoring								1	
Urban/regional planning	1								
Hydrology/water management									
Rangeland management									
Photogrammetry/cartography/ digital mapping			5	1	7	2	3		
Sensor development									
Hardware/software development									
Data management/information systems services									
Mathematics				1	1				

Total number of professional staff: 13   Total number of administrative staff: 6

Major Projects/Activities (1992-1996)

Title	Geographical extent/surface area in km²	Objectives	Generated products	Staff time	Names of other agencies involved in project/activity
Environmental degradation assessment of Mt. Pinatubo using ERS-1 SAR imagery	100,000	To map environmental degradation due to lahar	Digital maps, lahar data base, custom software	1 person-year	Philippine Institute of Volcanology and Seismology, University of Paris VI
Generation of input geographic databases to river flow models	100,000	To convert analogue thematic layers to digital format	Digital geographic databases; river flow model	1 person-year	PAGASA, INCEDE, National Hydraulic Research Centre

Major Current/Ongoing Projects/Activities

Title	Geographical extent/surface area in km²	Objectives	Generated products	Staff time	Names of other agencies involved in project/activity
Modelling siltation in Agno River	100,000	To model siltation processes in the Agno River	Digital maps, siltation models	0.5 person-year	National Hydraulics Research Centre, National Power Corporation

Title	Geographical extent/surface area in km^2	Objectives	Generated products	Staff time	Names of other agencies involved in project/activity
Utilization of AIRSAR data resource management activities	100,000	To extract from hyperspectral radar imagery of the Panay and Guimaras coasts		1 person-year	Department of Science and Technology
Lahar mapping using ADEOS AVNIR data	100,000	To model the lahar flow in 2 and 3 dimensions		1 person-year	
Fifty-year land-use change analysis of Ifugao Province	50,000	To determine long-term land-use change using aerial photos and satellite images		1 person-year	University of the Philippines Los Baños

Seminars/Conferences/Workshops Organized (1992-1996)

Type (seminar, conference, workshop, etc.)	Theme/title	Co-sponsorship (if any)	Number of participants	Title of output (publication and/or other)
Workshop	EC-ASEAN regional remote sensing workshop	ESA	20	
Workshop	National remote sensing workshop	ESA, Department of Science and Technology	40	
Workshop	GIS for health and environmental applications	Department of Health	50	
Workshop	Workshop on GIS	National Power Corporation	50	
Seminar	"3S" seminar for middle managers	AIT GIS Application Centre	30	

Training Activities (1992-1996)

Course title and description	Training undergone by organization staff		
	Given by (name of institution/agency)	Duration/ frequency	Number of staff taking course
Advanced seminar on remote sensing and GIS	AIT GIS Application Centre		1
GIS for landuse planning	ESCAP		1
Seminar on microwave remote sensing for oceanographic applications	EC, ASEAN, LAPAN		2
Seminar on microwave remote sensing for coastal applications	EC, ASEAN, MACRES		1

Course title and description	Training provided by organization for external users		
	Name of participating agencies/organizations	Duration/frequency	Number of participants
Introduction to remote sensing		3 per year	25
Introduction to GIS		3 per year	25
Applied remote sensing		3 per year	25
Advanced GIS		3 per year	25
Photo-interpretation		2 per year	10
Applied geodesy and photogrammetry		1 per year	10

Geodata Systems Technologies, Inc.

Geodata Systems Technologies, Inc., is the exclusive distributor of Radarsat and ESRI products in the Philippines.

Professional Personnel

Primary field of work	Number of professional staff with practical experience in			Number of staff with the following qualifications			Number of staff with intensive training (>2 months) during last 5 years in		
	RS	GIS	GPS	Ph.D.	MSc	BSc	RS	GIS	GPS
Agriculture/forestry/rangeland management									
Ecology/biology/biodiversity									
Geology/geomorphology soil sciences									
Geography						1			
Meteorology/climatology									
Oceanography and fisheries									
Environmental monitoring									
Urban/regional planning	1				1				
Hydrology/water management					2				
Rangeland management									
Photogrammetry/cartography/ digital mapping									
Sensor development									
Hardware/software development	2					2			
Data management/information systems services	6				1	5	1		

Total number of professional staff: 12   Total number of administrative staff: 7

Major Projects/Activities (1992-1996)

Title	Geographical extent/surface area in km²	Objectives	Generated products	Staff time	Names of other agencies involved in project/activity
Development of battleground simulation system study		To develop a battleground simulation system for military training activities using GIS			
GIS planning for Western Mining Corporation (WMC)		To develop a plan to effectively use GIS for the TAMPAKAN project			
Cagayan de Oro land-use planning project		To prepare a land-use city plan and conduct project feasibility study			
Fil-Estate lot information system		To design and instal a GIS-based lot information system for real estate			
GIS mapping of Philippine mining titles database		To generate GIS maps showing locations of 14,000 mining claims and titles			
Planning the aeromagnetic survey project using GIS		To organize spatial and other related map-based information for a survey project			
Mindanao development framework plan project		To provide GIS-based cartographic services			
Master plan study for West Central Luzon development programme		To provide technical advisory services in the inventory and assessment of spatial data			
Forest management support system feasibility study		To integrate remote sensing and GIS for forest resource management			
NAMRIA information technology strategic planning project		To develop such a plan for NAMRIA			
Assessment of preidentified forest lands for sustainable industrial forest plantation development and timber harvesting purposes	400	To undertake an inventory, determine the condition and make an assessment of timberland areas in Zamboanga del Sur			
Aerial photography and production of map-controlled photomosaic of timber lease agreement areas	16,000	To undertake aerial photography and construct photomosaic of timberland areas			
Aerial photo-interpretation and mapping of timber licence agreement areas		To assess the forest cover situation using GIS			
Development of GIS database for Clark Field, Pampanga		To develop a GIS database for land management and land-use planning			
Development of computerized environment impact assessment systems		To develop such a system for the Asian Development Bank			
Development of GIS database for Rizal Province		To develop a GIS database for land management and land-use planning			

Ongoing Projects/Activities

Title	Geographical extent/surface area in km²	Objectives	Generated products	Staff time	Names of other agencies involved in project/activity
Heritage Park GIS		To develop a GIS for burial plots, facilities and ongoing projects in Heritage Park			
AFP-JOC war room project		To design, develop and instal a multimedia GIS to support military data-gathering, analysis, management, decision-making, reporting and feedback			
GIS-enhanced information systems for pasig city goverment		To enhance existing information systems for local governance			
Iligan city comprehensive master plan development project		To prepare a city master development and land-use plan based on comprehensive city planning database built using GIS			

Seminars/Conferences/Workshops Organized (1992-1996)

Type (seminar, conference, workshop, etc.)	Theme/title	Co-sponsorship (if any)	Number of participants	Title of output (publication and/or other)
Seminar	Arc forest product	ESRI-Canada	60	
Seminar	ESRI-GIS users		75	
Conference	Radarsat	Radarsat	50	
Conference	Unix conference		47	
Seminar	IT 2000 LAB	NCC	30	
Workshop	Annual convention of the league of local planning and development coordination of the Philippines	DILG	70	
Seminar	UP-GIS	UP	30	

Databases/Archives (Including Remotely Sensed Data) Developed and/or Maintained

Type (digital or analogue) and storage medium	Database contents (type, location, source data, scale and other useful information)	Geographical extent/surface area in km²	Year of production/ review	Internet address (URL)	Main users
Digital, diskettes	Metro Manila from 1:10,000 NAMRIA maps. Information includes administrative boundaries, roads, landmark locations, rivers, etc. The data can be purchased and will be delivered in 12 diskettes. The file is in compressed form and instructions are included in the deliverables.	648	Started 1993- updated weekly		

Training Activities (1992-1996)

Course title and description	Training undergone by organization staff		
	Given by (name of institution/agency)	Duration/ frequency	Number of staff taking course
Surveying with global positiong system	NAMRIA	3 days	3

Course title and description	Training provided by organization for external users		
	Name of participating agencies/organizations	Duration/ frequency	Number of participants
Introduction to WS Arc Info		Quarterly	12 x 4 = 48
Introduction to PC Arc Info		Monthly	12 x 12 = 144
Introducing ArcView		Bi-monthly	12 x 2 x 12 = 288
Introducing ArcCad		Quarterly	10 x 4 = 40
Customizing Arc Info		Annual	12 x 1 = 12
Advanced Arc Info		Annual	12 x 1 = 12
Customizing Avenue with ArcView		Quarterly	12 x 4 = 48

Cybersoft Integrated Geoinformatics, Inc.

Cybersoft Integrated Geoinformatics, Inc., is a private institution involved in providing a total GIS solution to any spatial data and information management problems of private and government entities from data conversion, systems analysis and design to application and training.

Professional Personnel

Primary field of work	Number of professional staff with practical experience in			Number of staff with the following qualifications			Number of staff with intensive training (>2 months) during last 5 years in		
	RS	GIS	GPS	Ph.D.	MSc	BSc	RS	GIS	GPS
Agriculture/forestry/rangeland management	1	1							
Ecology/biology/biodiversity									
Geology/geomorphology/ soil sciences		1			1				
Geography									
Meteorology/climatology									
Oceanography and fisheries									
Environmental monitoring		1			1				
Urban/regional planning		2			2				
Hydrology/water management		1		1					
Rangeland management									
Photogrammetry/cartography/ digital mapping	1	20	1			2	2		
Sensor development									
Hardware/software development		2				5		2	
Data management/information systems services		2			1			3	
Physics		2		2					
Civil engineering						2			

Total number of professional staff: 61   Total number of administrative staff: 6

Major Projects/Activities (1992-1996)

Title	Geographical extent/surface area in km²	Objectives	Generated products	Staff time	Names of other agencies involved in project/activity
Geographic retail information and database system	294,000	To select sites for retail outlets, improve retailing and increase sales volume	Digital maps of road network and coast lines	20 person-days	
Geologic mapping	294,500	Ecological characterization	Digital maps	10 person-days	

Databases/Archives (Including Remotely Sensed Data) Developed and/or Maintained

Type (digital or analogue) and storage medium	Database contents (type, location, source data, scale and other useful information)	Geographical extent/surface area in km²	Year of production/ review	Internet address (URL)	Main users
Digital (8 mm tapes)	Road network and coastline of the Philippines (1:150,000)	294,500	1996		Private and government entities
Digital (8 mm tapes)	Urban cities in the Philippines (1:10,000; 1:5,000)	3,516	1996		Private and government entities
Digital (8 mm tapes)	25 per cent of all Philippine topographic maps (1:50,000)	74,000	1994/1996		Private and government entities

Training Activities (1992-1996)

Course title and description	Training undergone by organization staff		
	Given by (name of institution/agency)	Duration/ frequency	Number of staff taking course
ArcView training	Geodata Systems Technologies, Inc.	5 days	1
Genamap training	Genasys Malaysia	3 days	2
ER Mapper training	Earth Resource Mapping Pty. Ltd.	1 week	2

Course title and description	Training provided by organization for external users		
	Name of participating agencies/organizations	Duration/ frequency	Number of participants
Training on Atlas for Windows	Orient Integrated Development	5 days	5
Training on Atlas for Windows	National Economic Development Authority	2 days	10
Training on Atlas for DOS	Department of Health	5 days	5
Training on Atlas for DOS	Department of Environment and Natural Resources	5 days	5
Training on geographic retail information	Pilipinas Shell Corporation	1 week	20

National Water Resources Board

The National Water Resources Board is the body responsible for coordinating and integrating all activities related to water resources development and management. Its principal objective is to achieve a scientific and orderly development and management of all the water resources of the country with the principles of optimum utilization, conservation and protection to meet present and future needs.

P

Philippines

Professional Personnel

Primary field of work	Number of professional staff with practical experience in			Number of staff with the following qualifications			Number of staff with intensive training (>2 months) during last 5 years in		
	RS	GIS	GPS	Ph.D.	MSc	BSc	RS	GIS	GPS
Agriculture/forestry/rangeland management									
Ecology/biology/biodiversity									
Geology/geomorphology/ soil sciences									
Geography									
Meteorology/climatology									
Oceanography and fisheries									
Environmental monitoring									
Urban/regional planning									
Hydrology/water management		2			3	3			
Rangeland management									
Photogrammetry/cartography/ digital mapping									
Sensor development									
Hardware/software development									
Data management/information systems services		2						1	

Total number of professional staff: 6   Total number of administrative staff:

Major Projects/Activities (1992-1996)

Title	Geographical extent/surface area in km²	Objectives	Generated products	Staff time	Names of other agencies involved in project/activity
Philippine groundwater databank geographic information systems	Nationwide	To establish computerized national ground-water GIS and demonstrate how, and to what extent, the existing groundwater data, collected from different agancies can be utilized for rationalization of groundwater exploration, more accurate assessment of groundwater potential, planning of development, and protection and conserva-tion of groundwater resources	Standard well record, basic hydro-geologic maps	1 person-year	Local Water Utilities Administration, National Irrigation Administration, Metropolitan Waterworks and Sewerage System

Seminars/Conferences/Workshops Organized (1992-1996)

Type (seminar, conference, workshop, etc.)	Theme/title	Co-sponsorship (if any)	Number of participants	Title of output (publication and/or other)
Seminar	Water Code of the Philippines		80	
Seminar	Pricing policies and structures for urban and rural water supply in the ESCAP region	ESCAP	20	

Training Activities (1992-1996)

Course title and description	Training undergone by organization staff		
	Given by (name of institution/agency)	Duration/ frequency	Number of staff taking course
Philippine groundwater databanking	UNDP	2 years	3
Training on Autocad	Micro-circuit	2 weeks	1

Databases/Archives (Including Remotely Sensed Data) Developed and/or Maintained

Type (digital or analogue) and storage medium	Database contents (type, location, source data, scale and other useful information)	Geographical extent/surface area in km²	Year of production/ review	Internet address (URL)	Main users
Digital, diskettes	Groundwater and well data, pumping test data and water quality analysis	Nationwide	1993-present		Water agencies

P

4.2.14 Republic of Korea

Systems Engineering Research Institute

The Systems Engineering Research Institute (SERI) is a government-supported institution whose activities include development of software technology (computer systems, communication network, multimedia and image processing), construction of foundation for information-based society, super computer operation, national network construction and software engineering training. Its organizational chart is shown in figure 4.11.

Professional Personnel

Primary field of work	Number of professional staff with practical experience in			Number of staff with the following qualifications			Number of staff with intensive training (>2 months) during last 5 years in		
	RS	GIS	GPS	Ph.D.	MSc	BSc	RS	GIS	GPS
Agriculture/forestry/rangeland management									
Ecology/biology/biodiversity									
Geology/geomorphology/ soil sciences									
Geography									
Meteorology/climatology	5			3	2				
Oceanography and fisheries									
Environmental monitoring	4			1	3				
Urban/regional planning		7		2	5			2	
Hydrology/water management									
Rangeland management									
Photogrammetry/cartography/ digital mapping	1			1					
Sensor development									
Hardware/software development	1	3		1	3				
Data management/information systems services		3			3				

Total number of professional staff: 24   Total number of administrative staff: 3

Major Projects/Activities Completed (1992-1996)

Title	Geographical extent/surface area in km²	Objectives	Generated products	Staff time	Names of other agencies involved in project/activity
GIS development for environmental information management		To develop remote sensing and GIS application software for environmental monitoring		9 person-years	
Development of receiving and image processing systems for GMS/WEFAX using PCs		To develop GMS/WEFAX receiving and image processing systems		3 person-years	

Title	Geographical extent/surface area in km²	Objectives	Generated products	Staff time	Names of other agencies involved in project/activity
Development of image service system		To develop service system for satellite images on the information superhighway		3 person-years	
Study of comprehensive plan for implementing national GIS training programme		To implement GIS training programme at national level		4 person-years	Research institutes, universities, companies
Integrated system for water quality management				5 person-years	Chunghuk University Seoul National University
Development of advanced image processing techniques		To develop advanced image processing techniques, specially for detailed analysis of land surface information		3 person-years	
National GIS: system integration		To develop fundamental software for national GIS		4 person-years	

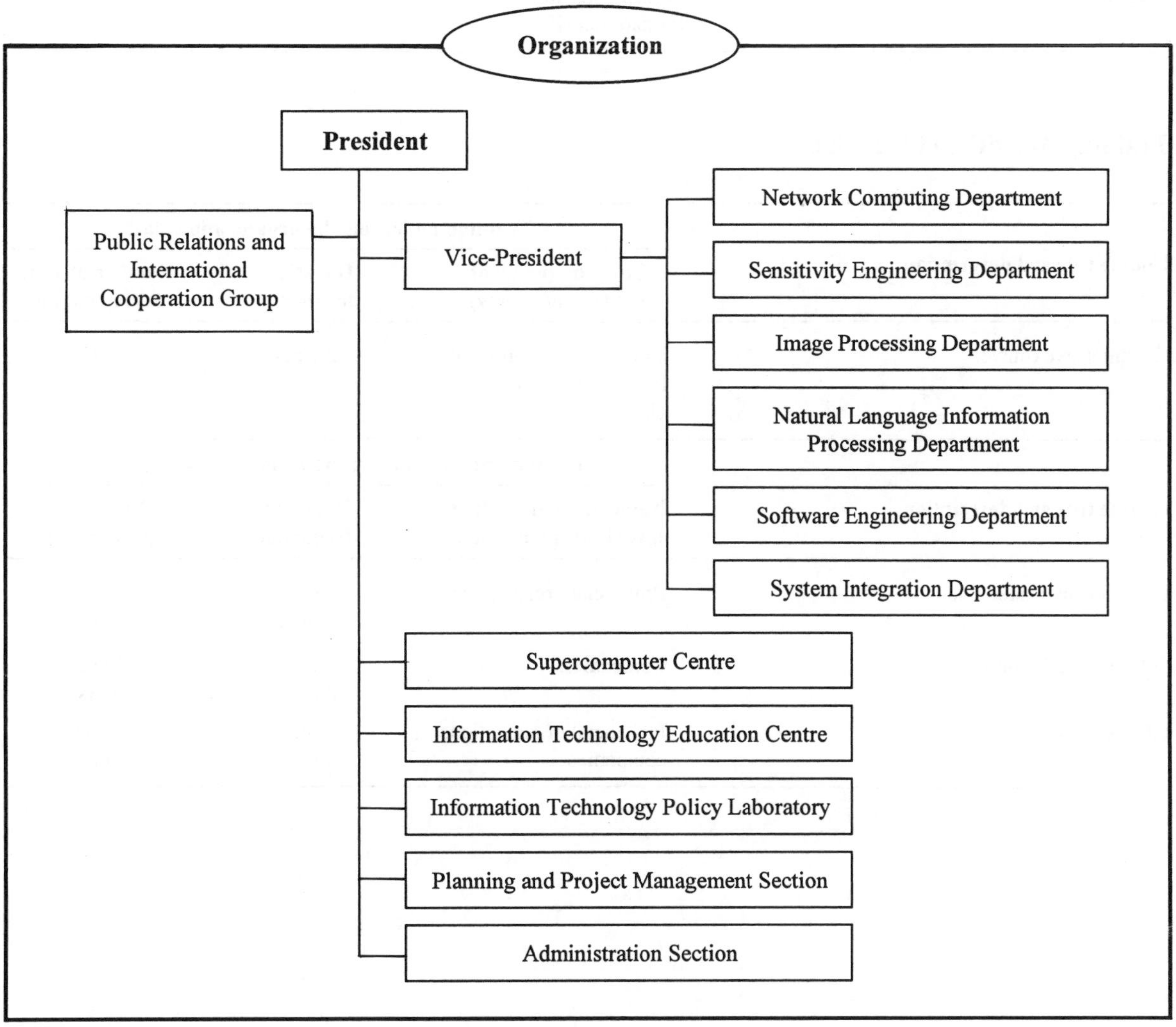

Figure 4.11 **Organizational chart of the Systems Engineering Research Institute**

183

Major Current/Ongoing Projects/Activities

Title	Geographical extent/surface area in km²	Objectives	Generated products	Staff time	Names of other agencies involved in project/activity
Development of techniques for analysis of remotely sensed images		To develop the software for thematic product generation, processing and data distribution		6.5 person-years	KAIST
Development of Web-based GIS		To develop interactive GIS software for the Web		4 person-years	
GIS editing software using Java language		To develop interactive GIS software for editing and integration		3 person-years	
Development of image service system		To further develop service system for satellite images on the information superhighway		3 person-years	
Development of advanced image processing techniques		To further develop advanced image processing techniques, specially for detailed analysis of land surface information		3 person-years	
National GIS: system integration		To further develop fundamental software for national GIS		4 person-years	

Training Activities (1992-1996)

Course title and description	Training undergone by organization staff		
	Given by (name of institution/agency)	Duration/ frequency	Number of staff taking course
GIS intensive course	University of California, Riverside, California, United States	2 weeks	2

Course title and description	Training provided by organization for external users		
	Name of participating agencies/organizations	Duration/ frequency	Number of participants
GIS decision maker's course	Provincial organizations	2 days/ 2 months	20 per course
GIS manager's course	Companies	2 weeks/ 2 months	20 per course
GIS users course	Provincial organizations/ companies	3 weeks/ 2 months	20 per course

4.2.15 Singapore

Centre for Remote Imaging, Sensing and Processing

The Centre for Remote Imaging, Sensing and Processing (CRISP) is a government organization mandated to develop remote sensing processing, imaging and analysis capabilities for applications in resource and environmental management; impart education and training in remote sensing data processing, analysis and applications; and receive, process, archive and disseminate satellite remote sensing data for research and operational uses.

Professional Personnel

Primary field of work	Number of professional staff with practical experience in			Number of staff with the following qualifications			Number of staff with intensive training (>2 months) during last 5 years in		
	RS	GIS	GPS	Ph.D.	MSc	BSc	RS	GIS	GPS
Agriculture/forestry/rangeland management	5	5			1				
Ecology/biology/biodiversity									
Geology/geomorphology/ soil sciences									
Geography									
Meteorology/climatology	2			2					
Oceanography and fisheries	4			1					
Environmental monitoring		1			1				
Urban/regional planning									
Hydrology/water management		1			1				
Rangeland management									
Photogrammetry/cartography/ digital mapping	3	3			2				
Sensor development									
Hardware/software development	5								
Data management/information systems services		3			3				
Physics				2	2				
Engineering/ground station						9			

Total number of professional staff: 25   Total number of administrative staff: 4

S

Major Current/Ongoing Projects/Activities

Title	Geographical extent/surface area in km²	Objectives	Generated products	Staff time	Names of other agencies involved in project/activity
Tropical forest study		Study of tropical forests with SAR data	Interfero-grams; classification charts		CESBIO, France
Ocean study		Study of internal waves, rain-cell signatures, etc.			Hamburg University
Ship detection and tracking			Software		Maritime and Port Authority
ERS LBR data applications		Assimilation of LBR data in numerical weather model			Meteoro-logical Services, Singapore
Red tide study		Assessment of water quality, detection of red tide	Algorithms		ESCAP, NASDA
Stereo-processing radargrammetry		Terrain computation			
SAR INSAR processing algorithms					

Other Information

Many seminars have been organized either solely by CRISP or jointly with SPOT Asia on SAR, visible remote sensing study of forests/vegetation, ocean phenomena, urban studies, etc., as well as remote sensing data processing techniques. Speakers have included Singaporean scientists as well as visiting scientists. CRISP staff are sent on various local or international training programmes on remote sensing, computer management, GIS, data processing techniques, etc. An example of local training is the workshop on Arc Info while an example of foreign training is the ESA Workshop on ERS Data Applications. CRISP has provided training on remote sensing and GIS for local and foreign trainees. CRISP has an archive catalogue-browse system on WWW (http://www.crisp.nus.sg) including quick look images of SPOT (about 40,000 scenes), ERS (about 8,000 scenes) and RADARSAT (newly developed). Various papers have been published in proceedings of conferences or research journals.

4.2.16 Sri Lanka

Environment and Forest Conservation Division, Mahaweli Authority

The Environment and Forest Conservation Division, Mahaweli Authority, is a semi-autonomous organization mandated to engage in regional environmental protection and conservation of the upper Mahaweli catchment.

Professional Personnel

Primary field of work	Number of professional staff with practical experience in			Number of staff with the following qualifications			Number of staff with intensive training (>2 months) during last 5 years in		
	RS	GIS	GPS	Ph.D.	MSc	BSc	RS	GIS	GPS
Agriculture/forestry/rangeland management	1					1			
Ecology/biology/biodiversity	1					1			
Geology/geomorphology/ soil sciences									
Geography									
Meteorology/climatology				1					
Oceanography and fisheries									
Environmental monitoring	1					1	1	1	
Urban/regional planning	1						1		
Hydrology/water management				1					
Rangeland management									
Photogrammetry/cartography/ digital mapping	2								
Sensor development									
Hardware/software development									
Data management/information systems services	1					1	1	1	

Total number of professional staff: 9   Total number of administrative staff:

S

187

Major Projects/Activities Completed (1992-1996)

Title	Geographical extent/surface area in km^2	Objectives	Generated products	Staff time	Names of other agencies involved in project/activity
Development of a GIS database for Koragahakande resources project	160	To assist in planning for Koragahakande resources project			
Upper Kotmale hydro project, inventory of land resources	605	To supply information needs for EIA for the upper Kotmale hydro project			Central Environmental Authority
Integrated map of land degradation in hilly regions of Sri Lanka	311	Identification of land degradation in UMC using GIS			Sri Lanka-Australia collaborative research
Collaborative programme with Tea Research Institute	176	Effective land-use planning in tea estates			Tea Research Institute
Preparation of resource inventory of VRR	460	Use GIS for preparing a resource inventory of VRR protected area			

Major Current/Ongoing Projects/Activities

Title	Geographical extent/surface area in km^2	Objectives	Generated products	Staff time	Names of other agencies involved in project/activity
Expansion of GIS database	4,000 sq. km.	Complete existing digital database			
Preparation of resources inventory of Mahaweli downstream areas	3,000 sq. km.	Assist the decision makers on issues concerning downstream areas			

Seminars/Conferences/Workshops Organized (1992-1996)

Type (seminar, conference, workshop, etc.)	Theme/title	Co-sponsorship (if any)	Number of participants	Title of output (publication and/or other)
GIS user's conference	To present GIS applications developed by members of the organization	Mahaweli Authority	55	
Indian remote sensing conference	Introduction of IRS data and practical application of remote sensing data in India	Mahaweli Authority	95	
GIS training for Namanukulu Plantations, Ltd.	GIS data management information for tea estate management	Mahaweli Authority	20	
GIS workshop	Natural resources management in South-East Asia	ODA	20	

Documents Published since 1990

Title	Year of publication	Language of publication	Keywords	Frequency
Geographical Information Systems for Natural Resources Management in South-East Asia	1994	English		
Impact of Land Degradation on Water Resources in the Hills of Sri Lanka	1994	English		

Training Activities (1992-1996)

Course title and description	Training undergone by organization staff		
	Given by (name of institution/agency)	Duration/ frequency	Number of staff taking course
Training in natural resources management	AIT	3 months	3
Cartography and GIS	Ordnance Survey	3 months	2
Training in GIS	Kingston University	3 months	1

S

4.2.17 Thailand

Thailand Remote Sensing Centre

The Thailand Remote Sensing Centre (TRSC) is a government organization mandated to operate the ground receiving station for satellite remote sensing data; acquire and process the satellite data; distribute remotely sensed data to users worldwide; develop remote sensing technology and facilities for image analysis; and promote transfer of remote sensing technology to users in the country and to provide researchers with funds. Its organizational chart is shown in figure 4.12.

Professional Personnel

Primary field of work	Number of professional staff with practical experience in			Number of staff with the following qualifications			Number of staff with intensive training (>2 months) during last 5 years in		
	RS	GIS	GPS	Ph.D.	MSc	BSc	RS	GIS	GPS
Agriculture/forestry/rangeland management	3	2	3	2	1				
Ecology/biology/biodiversity	1	1	1						
Geology/geomorphology/ soil sciences	1	1	1		1				
Geography	8	2	6	1	2	5			
Meteorology/climatology									
Oceanography and fisheries									
Environmental monitoring									
Urban/regional planning									
Hydrology/water management									
Rangeland management									
Photogrammetry/cartography/ digital mapping									
Sensor development									
Hardware/software development									
Operation and maintenance									
Data management/information systems services					1				

Total number of professional staff: 12   Total number of administrative staff: 3

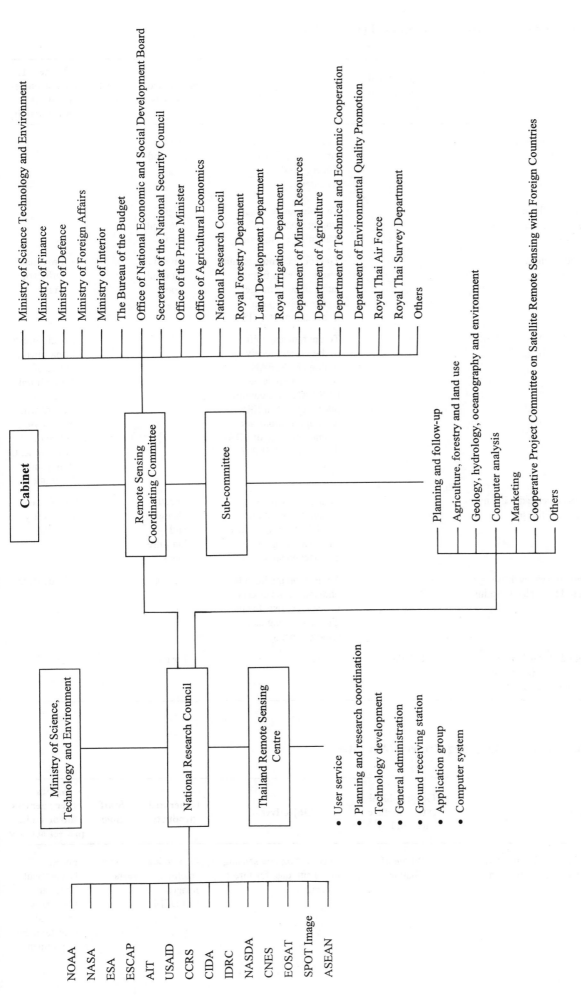

Figure 4.12 Organization chart of the Thailand Remote Sensing Centre

Major Projects/Activities Completed (1992-1996)

Title	Geographical extent/surface area in km²	Objectives	Generated products	Staff time	Names of other agencies involved in project/activity
Huai Kha Khaeng forest fire monitoring		To monitor and assess forest fire in Huai Kha Khaeng wildlife sanctuary	Digital map and report		
Model of flood area from construction of dam		To study and evaluate storage area of the Tha dam after construction	Digital data and report		
Remote sensing and GIS of Trang		To produce land-use and resources map from latest satellite imagery and compile spatial information into GIS	Digital data and report		Provincial Office of Trang
EC-ASEAN regional radar remote sensing ERS-1		To upgrade capability of the ground receiving stations in ASEAN countries to receive ERS SAR data, conduct pilot projects on ERS SAR applications and promote the use of ERS SAR data in ASEAN countries			Royal Forestry Department; Office of Agricultural Economics; Department of Fisheries; Office of Environmental Policy and Planning
Forest change detection in eastern forests		To assess changes in eastern forests during 1983-1993 and produce a land-use map of encroached area	Forest change and land-use maps		Department of Forestry
Study of sea level rise caused by global warming		To develop methods to understand the effects of sea level rise due to global warming using remote sensing	Report		GSI of Japan
Digital mosaicking of NOAA AVHRR imagery of South-East Asia	8,700,000		Poster		

Major Current/Ongoing Projects/Activities

Title	Geographical extent/surface area in km²	Objectives	Generated products	Staff time	Names of other agencies involved in project/activity
Global research network system (negotiation data sets)	Whole of Thailand	To collect remote sensing and in situ data for forest coverage	Geo-coded satellite images and DEM	5 person-years	Forest Department, Kasetsart University and Land Development Department

Title	Geographical extent/surface area in km²	Objectives	Generated products	Staff time	Names of other agencies involved in project/activity
IGBP land-use/land-cover change	Whole of Thailand	To monitor land-use/land-cover changes in tropical forest area using various types of remote sensing data (1980-1994), identify and analyse the factors causing the changes and their impact on the environment and develop a dynamic model for monitoring land-use/land-cover changes in tropical forest areas and to apply the model to assess future trends up to 2020		3 person-years	Forest Department, Land Development Department and Chiang Mai University
GlobeSAR Thailand	Whole of Thailand	To evaluate and promote use of SAR data (airborne and Radarsat)		5 person-years	Forest Department, Land Development Department, Khon Kaen University and Suranaree University of Technology
AIRSAR Thailand	Whole of Thailand	To evaluate and promote multi-band airborne SAR data		5 person-years	Departments of Forest, Land Development, Thai Survey, Fisheries and Naval Hydrography
Digital mosaicking of Landsat TM imagery of Thailand	Whole of Thailand		Digital	1-2 person-years	

Seminars/Conferences/Workshops Organized (1992-1996)

Type (seminar, conference, workshop, etc.)	Theme/title	Co-sponsorship (if any)	Number of participants	Title of output (publication and/or other)
Workshop	SAR data analysis and applications	RESTEC/STA (Japan)	120	Proceedings
Seminar	EC-ASEAN national seminar for decision makers	AIT, CEC, ESA, UNDP, UNEP GRID	205	
Workshop	EC-ASEAN PIs and CIs workshop	European Union, ESA	24	
Workshop	1st GlobeSAR international workshop	CCRS, IDRC	100	Proceedings

T

Documents Published since 1990

Title	Year of publication	Language of publication	Keywords	Frequency
TRSC Newsletter	1982-present	Thai/English		Quarterly
Proceedings of workshop on SAR data analysis and applications	1993	English		
EC-ASEAN project report	1997	Thai		
Manual of ERS-SAR applications	1997	Thai		

Training Activities (1992-1996)

Course title and description	Training undergone by organization staff		
	Given by (name of institution/agency)	Duration/ frequency	Number of staff taking course
ESA-United Nations training course on ERS data	ESA	2 weeks	1
ESA training on application of microwave remote sensing to disasters	ESA	2 weeks	2

Course title and description	Training provided by organization for external users		
	Name of participating agencies/organizations	Duration/ frequency	Number of participants
EC-ASEAN regional training	Relevant agencies from Thailand, Indonesia, Malaysia, Philippines	2 weeks	24
EC-ASEAN national training on radar applications to natural resources surveying	Government agencies of Thailand	2 weeks	24
Hands-on training on digital image processing	Government and private sector agencies of Thailand	2 weeks	20-25

Databases/Archives Developed and/or Maintained and Available to Outside Users

Type (digital or analogue) and storage medium	Database contents (type, location, source data, scale and other useful information)	Geographical extent/surface area in km²	Year of production/ review	Internet address (URL)	Main users
Digital, Exabyte	Geocoded satellite (Landsat and SPOT) images: Chiang Mai, Kanchanaburi, Phang Nga	Chiang Mai, Kanchanaburi, Phang Nga	1996		
Digital, Exabyte	DEM: Chiang Mai, Phang Nga, Kanchanaburi	Chiang Mai, Kanchanaburi, Phang Nga	1996		
Digital, Homepage	Geocoded satellite images: Huai Kha Khaeng Wildlife Sanctuary	Huai Kha Khaeng Wildlife Sanctuary	1995		
Digital, Homepage	Spatial database of Trang Province	Trang Province	1994		
Digital, Homepage	Model of dam geocoded satellite images: Thadan Village, Nakhon Nayok Province		1994		

4.2.18 Viet Nam

National Centre for Science and Technology of Viet Nam

The National Centre for Science and Technology of Viet Nam is a government organization engaged in developing national policy on science and technology, conducting main national scientific projects, and coordinating scientific activities in the country.

Professional Personnel

Primary field of work	Number of professional staff with practical experience in			Number of staff with the following qualifications			Number of staff with intensive training (>2 months) during last 5 years in		
	RS	GIS	GPS	Ph.D.	MSc	BSc	RS	GIS	GPS
Agriculture/forestry/rangeland management									
Ecology/biology/biodiversity	2			2		1	1		
Geology/geomorphology/ soil sciences	4			2		2	2		
Geography	3	2		1	2	2	3	1	
Meteorology/climatology									
Oceanography and fisheries	4					4	2		
Environmental monitoring	10	2		1	1	10	4	2	
Urban/regional planning									
Hydrology/water management	2			1		1	1		
Rangeland management									
Photogrammetry/cartography/ digital mapping	6	3		2	6	1	5	2	
Sensor development									
Hardware/software development	9	5		2	4	8	5		
Data management/information systems services									

Total number of professional staff: 52   Total number of administrative staff:

Major Projects/Activities Completed (1992-1996)

Title	Geographical extent/surface area in km²	Objectives	Generated products	Staff time	Names of other agencies involved in project/activity
Application of remote sensing to land-use mapping	250,000	Preparation of land-use map of whole country at scale of 1:250,000	Land-use map (1:250,000 scale)	20 person-years	Department of Geodesy and Cartography; Institute of Forest

V

Major Current/Ongoing Projects/Activities

Title	Geographical extent/surface area in km^2	Objectives	Generated products	Staff time	Names of other agencies involved in project/activity
Use of SAR imagery for flood monitoring	100,000	Mapping of the flooding in the Mekong River delta	Map of floods	5 person-years	Institute of Meteorology and Hydrology

Documents Published since 1990

Title	Year of publication	Language of publication	Keywords	Frequency
Proceedings of the National Centre for Science and Technology		English		Every 3 months

4.3 Ground receiving stations: overview and country profiles

The ground segment has kept pace with the advances in the space segment as more sophisticated and versatile image processing and analysis and GIS software and more powerful computers continue to be developed and marketed. As a result of improved pre-processing capabilities, ground receiving stations now offer data of much superior quality and a wider range of data product types, formats and media, including precisely geo-referenced data sets, merged images and other enhanced images. Similarly, user agencies can extract more information from satellite data and produce various kinds of value-added derived products, like thematic maps, digital elevation/terrain models (DEM/DTM) and 3-D perspectives, applying powerful image processing software. A large number of countries in the Asian and Pacific region now have sophisticated facilities for digital processing and analysis of satellite remote sensing data as well as versatile GIS facilities. Such facilities are now installed not only in the main national space and remote sensing centres (which was the case until a few years ago), but in an increasing number of end-user agencies as well. Remote sensing data from the various currently operational satellites are thus being applied, often in conjunction with GIS, for a wide variety of studies relating to resource and environmental surveying, mapping and management covering a range of Earth science disciplines, as mentioned in subsection 4.1.1 above. The use of satellite-based positioning techniques to complement remote sensing and GIS in surveying and mapping applications has also been gradually increasing in recent years.

The ESCAP region has as many as 11 ground stations for the reception, processing, archiving and dissemination of data from various remote sensing satellites. There are two stations each in Australia (Alice Springs and Hobart) and Japan (Hatoyama and Kumamoto), and one station each in China (Beijing), India (Hyderabad), Indonesia (Pare Pare), Pakistan (Islamabad), the Republic of Korea (Taejon), Singapore and Thailand (Bangkok). At present, these stations are collectively acquiring data from eight different dedicated remote sensing satellite systems, namely Landsat, SPOT, MOS, IRS, ERS, JERS, ADEOS and Radarsat. In addition, data from what are categorized essentially as meteorological satellites, such as NOAA and GMS, are also received at some of these stations. The region has the highest concentration of such ground receiving stations in the world, and some areas in South, South-East and North-East Asia fall in the coverage zones of three or even four stations.

4.3.1 Australia

Ground Receiving Station for Satellite Remote Sensing Data in Australia

1. **General information**

Located at/near:	(1) Hobart
	(2) Alice Springs
Geographic coordinates:	(1) 42°53'S, 147°19'E
	(2) 23°41'S, 133°52'E
Agency responsible for operating station:	ACRES
Year of establishment:	
Acquisition zone:	See map at figure 4.13
Data being received:	(1) ERS, JERS, SPOT
	(2) AVHRR, ERS, LERS, Landsat, Radarsat, SPOT

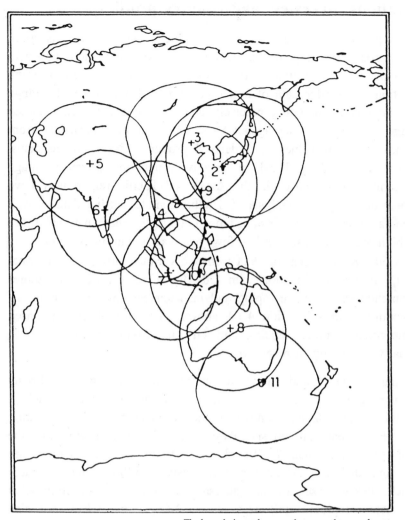

Legend:

1. Hatoyama (Japan)
2. Kumamoto (Japan)
3. Beijing (China)
4. Bangkok (Thailand)
5. Islamabad (Pakistan)
6. Hyderabad (India)
7. Jakarta (Indonesia)
8. Alice Springs (Australia)
9. Taiwan Province of China
10. Parepare (Indonesia)
11. Hobart (Australia)

Note: Singapore not shown.

Figure 4.13 Coverage areas of Asia-Pacific high-resolution ground stations

4.3.2 China

Ground Receiving Station for Satellite Remote Sensing Data in China

1.   **General information**

    Located at/near:

    Geographic coordinates: 40°27' 6.07"N, 116°51' 27.55"E

    **Agency responsible for
operating station:**   1986

    Year of establishment:

    Acquisition zone: See map at figure 4.13

    Data being received:

Landsat (MSS/TM)	since 1986
SPOT (XS/Pan)	since 1997
ERS-1	since 1995
JERS-1	since 1994
Radarsat	since 1997

2. **Data acquisition facilities**

   Antenna systems: SA 10-m

   Recording media: HDDT, DLT

   Tracking and telemetry equipment: SA 10-m

   Other equipment:

3. **Data pre-processing and photo-processing facilities**

   Computer systems: VAX 6250

   Array processors: Star-VD1

   Workstations: SUN, Silicon Graphics, Octane

   Pre-processing/processing software: STX, MDA, Meridian

   Photo-processing facilities:

4. **Data product generation**

   Digital products

   Format:

   Media: CCT, HDDT, DLT

   Photographic products: B/W and colour prints

   Other/special products

   Digital:

   Photographic:

   Throughput capacity: Landsat (16/day), ERS (6/day), JERS (6/day)

5. **Data archiving and distribution**

   Digital storage media: CCT, HDDT, DLT

   Size of data archive:

   Data cataloguing type: Computerized browse catalogue; paper prints

   Data distributed/supplied through: Physical transfer through airmail/courier

   Availability of on-line electronic
   services on Internet or other
   international network: http://www.rsgs.ac.cn

   Networks
   Local area network (LAN): Yes
   Wide area network (WAN): Yes

4.3.3 Indonesia

Ground Receiving Station for Satellite Remote Sensing Data in Indonesia

1. **General information**

   Located at/near:

   Geographic coordinates: 3°55'S, 119°35'E

   Agency responsible for
   operating station: LAPAN

   Year of establishment:

   Acquisition zone: See map at figure 4.13

Data being received:	Landsat (MSS/TM)	since 1993
	SPOT (XS/Pan)	since 1993
	ERS-1	since 1993
	JERS-1	since 1995
	NOAA	since 1981
	GMS	since 1982

2. **Data acquisition facilities**

Antenna systems:	X-band (2), S-band Cassegrain (4) L-band (2)
Recording media:	DCRSi (4), CCT (1)
Tracking and telemetry equipment:	For Landsat, SPOT, ERS-1 (2)
	For NOAA, GMS (4)
Other equipment:	

3. **Data pre-processing and photo-processing facilities**

Computer systems:	DEC 4000 (2)
Array processors:	Star Array Processor VP Series (3)
Workstations:	Meridian (4)
Pre-processing/processing software:	GICS/MDA, METDAS/NOAA, DUNDEE/GMS
Photo-processing facilities:	Kodak

4. **Data product generation**

Digital products

Format:	LOGSWG, CIRS
Media:	CCT, CD-ROM, EXABYTE
Photographic products:	Hard copy

Other/special products

Digital:	
Photographic:	
Throughput capacity:	Film/hard copy (10), Digital (12)

5. **Data archiving and distribution**

Digital storage media:	DCRSI, CCT, CD-ROM, EXABYTE
Size of data archive:	
Data cataloguing type:	Browse-catalogue
Data distributed/supplied through:	Physical transfer through mail/courier
Availability of on-line electronic services on Internet or other international network:	Not applicable

Networks

Local area network (LAN):	No
Wide area network (WAN):	No

4.3.4 Japan

Ground Receiving Station for Satellite Remote Sensing Data in Japan

1. **General information**

   Located at/near:

   Geographic coordinates: 36°0'N, 139°22'E

   Agency responsible for
   operating station: NASDA

   Year of establishment: 1978

   Acquisition zone:

   Data being received from:

Landsat (MSS/TM)	since 1979
SPOT (XS/Pan)	since 1988
MOS	1987-1996
ERS-1	1991-1996
JERS-1	since 1992
ADEOS	since 1996

2. **Data acquisition facilities**

   Antenna systems: 10 metre (2) and 11.5 metre (1) for S and X-bands

   Recording media: HDDT (6), D-1 (16)

   Tracking and telemetry equipment: Satellite-specific telemetry

   Other equipment:

3. **Data pre-processing and photo-processing facilities**

   Computer systems: 1 set each for (i) Landsat, SPOT, MOS, (ii) JERS-1, ERS-1, and (iii) ADEOS

   Array processors:

   Workstations: Many

   Pre-processing/processing software: Original

   Photo-processing facilities:

4. **Data product generation**

   Digital products

   Format: CEOS, HDF, SKIMPY

   Media: CCT, D-1 cassette, 8 mm DAT, CD-ROM, floppy

   Photographic products: B and W and colour positive/negative films and prints

   Other/special products

   Digital:

   Photographic:

   Throughput capacity: Landsat (2), SPOT (21), JERS-1 OPS (4.5), JERS-1 SAR (3.5), ADEOS AVNIR (4), ADEOS OCTS (1-8, depending on product)

5. **Data archiving and distribution**

   Digital storage media: D-1, HDDT, 8 mm, CCT, floppy

   Size of data archive: Large amounts of data from the various satellites listed against item 6 under General Information above have been archived at the station

Data cataloguing type:	Browse
Data distributed/supplied through:	Physical transfer through express mail/courier
Availability of on-line electronic services on Internet or other international network:	Yes
Internet address:	www.eoc.nasda.go.jp
Address on other network:	
Networks	
Local area network (LAN):	Ethernet
Wide area network (WAN):	Private line

4.3.5 Pakistan

Ground Receiving Station for Statellite Remote Sensing Data in Pakistan

1. General information

Located at/near:	Islamabad	
Geographic coordinates:	33°31'N, 73°11'E	
Agency responsible for operating station:	SUPARCO	
Year of establishment:	1989	
Acquisition zone:	See map at figure 4.13	
Data being received from:	Landsat (MSS/TM)	since 1989
	SPOT (XS/Pan)	since 1989
	NOAA AVHRR HRPT	since 1989

2. Data acquisition facilities

Antenna systems:	(i) Cassegrain dual feed 10 metre for Landsat and SPOT
	(ii) Cassegrain dual feed 2.5 metre for NOAA (1)
Recording media:	HDDT (3), RAIDs (8), DLT 7000 (4) for Landsat/SPOT CCT for NOAA
Tracking and telemetry equipment:	S, X Band receivers (2)
Other equipment:	ENERTIC chain for Landsat, SPOT NOAA chain for NOAA

3. Data pre-processing and photo-processing facilities

Computer systems:	DEC VAX 11/780 (2)
	DEC AlphaServer 4000 5/300 (biprocessor) (2)
Array processors:	FPS 5210 (4)
Workstations:	I²S (4)
	Graphics workstation (2)
Pre-processing/processing software:	(i) Image processing software developed by SEP
	(ii) SatINGEST, SatView, SatSTORE
	(iii) Landsat and SPOT processing software provided by MATRA
Photo-processing facilities:	Enlargers, developers, contact printer, Vizirmatic laser film recorders

4. Data product generation

Digital products

Format:	SISA for SPOT, CCRS and ESA for Landsat, NOAA 1B for NOAA
Media:	CCT, CD-ROM, EXABYTE for Landsat and SPOT and CCT for NOAA
Photographic products	B and W and colour positive/negative films (240 mm) B and W and colour prints at various scales
Other/special products	
Digital:	
Photographic:	
Throughput capacity:	Landsat (45), SPOT (200), NOAA (4)

5. **Data archiving and distribution**

Digital storage media:	HDDT, DLT for Landsat and SPOT, CCT for NOAA
Size of data archive:	Landsat TM (37,000 scenes) Landsat MSS (5,400 scenes) SPOT (25,000 scenes)
Data cataloguing type:	Customer Info System RDB/VMS (SEP); Catalogue Browse System (MATRA)
Data distributed/supplied through:	Physical transfer through mail/courier
Availability of on-line electronic services on Internet or other international network:	No
Internet address: **Address on other network:**	
Networks	
Local area network (LAN):	No
Wide area network (WAN):	No

4.3.6 Singapore

Ground Receiving Station for Satellite Remote Sensing Data in Singapore

1. **General information**

Located at/near:	Singapore	
Geographic coordinates:	1°18′N, 103°47′E	
Agency responsible for operating station:	Centre for Remote Imaging, Sensing and Processing (CRISP), National University of Singapore	
Year of establishment:	1995	
Acquisition zone:	See map at figure 4.13	
Data being received from:	SPOT	since September 1995
	ERS	since March 1996
	Radarsat	since December 1996

2. **Data acquisition facilities**

Antenna systems:	Datron – 13 metre diameter for X-Band
Recording media:	SONY DIR 1000 HDDRs (2)
Tracking and telemetry equipment:	Datron
Other equipment:	

3. **Data pre-processing and photo-processing facilities**

    Computer systems: Silicon Graphics Challenge

    Array processors:

    Workstations: SGI INDIGO (4)

    Pre-processing/processing software: Datron

    Photo-processing facilities:

4. **Data product generation**

    Digital products:

    Format: CEOS, SPOT CAP, SPOTView

    Media: CCT, CD-ROM, EXABYTE

    Photographic products:

    Other/special products

    Digital:

    Photographic:

    Throughput capacity: SPOT (20), SAR (6)

5. **Data archiving and distribution**

    Digital storage media: SONY, DIR 1000 HDDRs (2)

    Size of data archive:

    Data cataloguing type: World Wide Web

    Data distributed/supplied through:

    Availability of on-line electronic
    services on Internet or other
    international network: Yes

    Internet address: http://www.crisp.nus.sg
    Address on other network:

    Networks
    Local area network (LAN):   Ethernet
    Wide area network (WAN):   No

4.3.7 Thailand

Ground Receiving Station for Satellite Remote Sensing Data in Thailand

1. **General information**

    Located at/near: Bangkok

    Geographic coordinates: 13°31'N, 100°11'E

    Agency responsible for
    operating station: Thailand Remote Sensing Centre

    Year of establishment: 1982

    Acquisition zone: See map at figure 4.13

    Data being received from: Landsat (MSS/TM)   since 1982
                                            SPOT (XS/Pan)   since 1987
                                            MOS   1988-1996
                                            ERS-1   since 1992
                                            JERS-1   since 1994
                                            MOAA AVHRR HRPT   since 1992

204

There were plans to receive IRS, Radarsat and ADEOS data from the end of 1997.

2. **Data acquisition facilities**

Antenna systems:	For S and X band
Recording media:	HDDT
Tracking and telemetry equipment:	Various
Other equipment:	

3. **Data pre-processing and photo-processing facilities**

Computer systems:	Various
Array processors:	Star-VP1
Workstations:	Various
Pre-processing/processing software:	
Photo-processing facilities:	Colour Fire 2240

4. **Data product generation**

Digital products:

Format:	Various (CCRS, LGSOWG)
Media:	CCT, EXABYTE
Photographic products:	240 mm film and prints
Other/special products	
Digital:	
Photographic:	
Throughput capacity:	Landsat (45), SPOT (200), NOAA (4)

5. **Data archiving and distribution**

Digital storage media:	HDDT, DLT for Landsat and SPOT, CCT for NOAA
Size of data archive:	Large amounts of data from the various satellites listed against item 6 under General Information above have been archived at the station.
Data cataloguing type:	Hard copy (microfiche)
Data distributed/supplied through:	Physical transfer through mail/courier
Availability of on-line electronic services on Internet or other international network:	Yes
Internet address:	http://www.nrct.go.th
Address on other network:	
Networks	
Digital storage media:	Yes
Wide area network (WAN):	No

Chapter 5

SATELLITE METEOROLOGY AND DISASTER MONITORING

5.1 Regional overview

5.1.1 Background

The photographs taken by the early United States Gemini missions of the late 1950s and early 1960s, though crude by today's standards, indicated the feasibility of using satellites to obtain, from their vantage points in space, synoptic and repetitive images of the Earth for monitoring weather systems. Accordingly, the United States launched the first of the operational polar-orbiting NOAA/TIROS series of weather satellites in the early 1960s. Satellite meteorology was thus one of the earliest applications of space research to acquire an operational or semi-operational status. Because of their huge areal coverage and the repetitiveness of their observations, meteorological satellites were ideally suited for monitoring and studying the movement of clouds, storms, cyclones, typhoons and hurricanes which typically range over huge distances. As more satellites in the operational NOAA series were launched, improvements in the ground segment, in terms of receiving and processing equipment and data analysis techniques, continued to be made. Satellite meteorology came to be increasingly employed for routine weather monitoring and forecasting. Each NOAA satellite provides twice daily coverage of every part of the Earth, and with a pair of synchronized satellites in orbit, as is almost always the case, repeat coverage four times a day is assured, which makes it possible for many dynamic events to be quite closely monitored. The former Soviet Union also launched its polar orbiting Meteor series of operational meteorological satellites with roughly comparable weather monitoring capabilities.

The 1970s witnessed the launch of various meteorological and environmental satellite systems in the geostationary orbit. Geostationary meteorological satellites are capable of providing images with much higher observational frequencies, i.e. every hour or half-hour, enabling even rapidly changing events to be closely monitored. Examples of such geostationary meteorological and environmental satellite systems launched in the 1970s were the United States' GOES, the ESA Meteosat and Japan's GMS (Himawari). If data from polar orbiting satellites that typically have an altitude of about 800-900 km, and from geostationary satellites, which are at an altitude of about 36,000 km, are available, they can be gainfully used in a complementary manner. Over the decades, satellite data from the dedicated, operational meteorological satellites, whether in the polar or geostationary orbits, have become an integral and indispensable part of weather forecasting and disaster monitoring activities.

5.1.2 Regional meteorological satellite systems

In addition to the NOAA series, which cover the whole world, the GOES West, which covers the Pacific region, and ESA Meteosat satellites, which cover Western Asia, the ESCAP region has several satellite systems operated by regional countries. All four spacefaring countries of the region, namely China, India, Japan and the Russian Federation, have launched their own operational meteorological satellites. The Russians have their polar-orbiting Meteor and geostationary GOMS. China has its polar-orbiting FY-1 and geostationary FY-2 satellites. Japan has its GMS series of geostationary meteorological satellites. India has its INSAT series of satellites, which are multi-purpose meteorological-cum-communication satellites in the geostationary orbit.

5.1.3 Use of meteorological satellite data in the region

The Asian and Pacific region collectively has a large number of ground stations for accessing and processing data from the various meteorological satellites. The total number of such installations in the region would be of the order of a few hundred. In most cases, data are received in the APT mode with a degraded spatial resolution, which, nevertheless, is sufficient for routine weather forecasting purposes. There are also facilities in some stations for reception of HRPT data at full spatial resolution. These data, especially HRPT data, are additionally used for various non-meteorological purposes, such as monitoring the vegetation index, which gives an indication of vegetation/crop vigour; studying sea surface temperatures, which are important for oceanography studies as well as fishing; and locating areas that may be suitable for siting solar power generation facilities, on account of their receiving much solar insulation because of low levels of cloudiness during the year. In addition, NOAA has other on-board sensors, like the TOVS, which provides vertical profiles of a host of atmospheric parameters like temperature, humidity, dewpoint, pressure, wind speed/direction and total ozone content from the ground to a height of about 35 km. TOVS provides invaluable data sets for research on various tropospheric and stratospheric processes, as well as for weather forecasting. Receiving stations for NOAA and other meteorological satellites are usually located in national meteorological departments or bureaux for operational weather and natural hazard forecasting, and/or in national space agencies, where they are applied for meteorological, climatological or atmospheric research. Much expertise in the application of meteorological satellite data has thus been built up in a large number of countries in the region.

The region has, by and large, made good use of the potential of meteorological satellites for weather forecasting and disaster monitoring. The region is particularly prone to natural disasters like typhoons, cyclones, floods, landslides, tsunamis, tidal surges and volcanic eruptions, which have been causing devastation on a massive scale with huge loss of life. Over 50 per cent of the world's natural disasters hit the Asian and Pacific region. Therefore, the use of satellite data for weather forecasting and disaster monitoring, assessment, preparedness and relief is very relevant in the context of this region. As a result of the increasing and sustained operational use of space technology for weather and disaster monitoring in the region, which enables advance warnings and alerts about impending natural disasters to be issued and timely evacuations to be made from areas about to be struck by the disasters, many more precious human lives are now being saved than was possible only a few years ago. In Bangladesh, in particular, there has been a dramatic reduction in the number of fatalities resulting from cyclones and storm surges, as a result of more accurate weather forecasts, timely warnings and evacuations, all of which have been made possible by space technology.

In view of the relevance and importance of satellite meteorology and disaster monitoring to the region, this field has been selected as one of the four "core" application areas for the Regional Space Applications Programme for Sustainable Development (RESAP). Accordingly, a Regional Working Group on Meteorological Satellite Applications and Natural Hazards Monitoring was established under RESAP to promote use of space technology in meteorology, disaster monitoring and related applications in regional countries. At the second meeting of this working group, held in April 1997 in Phuket, Thailand, three priority areas were identified, namely monsoon monitoring, natural hazards monitoring and assessment, and non-meteorological applications. It is expected that participating countries will undertake meaningful collaborative projects and activities that could lead to practical and concrete benefits. The names of the national contact points for the Regional Working Group are provided in table 5.1.

5.1.4 Conclusion

The Asian and Pacific region, as a whole, is well equipped to exploit the potential of meteorological and environmental satellites for weather forecasting, disaster monitoring, atmospheric and climatological research and other non-meteorological uses, like monitoring of vegetation index and sea surface temperatures. The region has its own complement of both polar orbiting and geostationary meteorological satellites and can also access satellite systems operated by extra-regional entities. It has, by and large, a fairly good infrastructure on the ground, in the form of facilities for the reception, processing, analysis and application of meteorological satellite data, as well as trained and experienced manpower. It is hoped that the application of data from the meteorological and environmental satellites for environmental management, disaster mitigation and atmospheric research will contribute appreciably to the overall goal of sustainable development in the region.

Table 5.1 National contact points of the Regional Working Group on Meteorological Satellite Applications and Natural Hazards Monitoring

Country	Name	Agency
Australia	Mr A.B. Neal	Department of Environment, Sport and Territories
Azerbaijan	–	–
Bangladesh	Mr A.M. Choudhury	Space Research and Remote Sensing Organization
China	Professor Xu Jianmin	Centre for Meteorological Satellites, CMA
Fiji	Mr Rajendra Prasad	Ministry of Tourism and Civil Aviation
India	–	–
Indonesia	Mr Sujadi Hardjawinata	Meteorological and Geophysical Agency
Islamic Republic of Iran	Ms Mehrnaz Bijianzadeh	Iranian Meteorological Organization
Japan	Mr Masaro Saiki	Japan Meteorological Agency
Malaysia	Mr Lim Joo Tick	Malaysian Meteorological Services
Mongolia	Ms M. Erdene-tuya	Agency of Meteorology and Environment Monitoring
Myanmar	–	–
Nepal	Mr Kiran Shankar Yogacharya	Department of Hydrology and Meteorology
Pakistan	Mr Z.R. Siddiqui	Space and Upper Atmosphere Research Commission
Philippines	Mr Cipriano Ferraris	Philippine Atmospheric, Geophysical and Astronomical Services Administration
Republic of Korea	–	–
Russian Federation	–	–
Singapore	–	–
Sri Lanka	Director	Department of Meteorology
Thailand	Mr Dusadee Sarigabutr	Meteorological Department
Vanuatu	–	–
Viet Nam	Mr Hoang Minh Hien	National Centre for Hydrometeorological Forecasting

Table 5.2 Existing satellite meteorology facilities/infrastructure (hardware)

Country	Agency	Computers			Network	Radio equipment	Other equipment
		MF	WS	PC			
Azerbaijan	Azerbaijan National Space Agency	✗	✗	✓	LAN		
Bangladesh	Space Research and Remote Sensing Organization	✓	✗	✓			Ground station
China	National Satellite Meteorology Centre	✓	✓	✓	LAN		
	Institute of Remote Sensing Applications	✗	✓	✓	Internet		
	Wuhan Technical University of Surveying and Mapping	✗	✓	✓	CERNET		
Indonesia	National Aeronautics and Space Institute	✓	✓	✓			
	Meteorological and Geophysical Agency	✓	✓	✗	LAN, Internet		
Islamic Republic of Iran	Islamic Republic of Iran Meteorological Organization	✓	✓	✓	Internet	Radio transmitter	Ground station
Japan	Japan Meteorological Agency	✓	✓	✓	LAN, Internet	Radio and microwave equipment	
Malaysia	Malaysian Meteorological Service	✗	✓	✓			
Mongolia	National Remote Sensing Centre	✓	✓	✓	LAN, WAN, Internet	TV channel	
Pakistan	Space and Upper Atmosphere Research Commission	✗	✓	✓	DCP network	Fax/modem	
Philippines	Philippine Atmospheric, Geophysical and Astronomical Services Administration		✓	✓	LAN		Ground station
Republic of Korea	Korean Meteorological Research Institute	✓	✗	✗			
Singapore	Meteorological Service Singapore	✓	✓	✓	LAN		
Sri Lanka	Department of Meteorology	✗	✗	✓	LAN		
Viet Nam	National Centre for Hydrometeorological Forecasting	✗	✓	✓			Leased line-satellite circuit

Table 5.3 Existing satellite meteorology facilities/infrastructure (software)

Country	Agency	Software	Database
Azerbaijan	Azerbaijan National Space Agency	Self-developed	
Bangladesh	Space Research and Remote Sensing Organization	I^{2}S-IVAS System 600, GMS-5 processing package	
China	Institute of Remote Sensing Applications	Erdas, ER Mapper, PCI, Envi, Arc Info, IRSA-II	

Table 5.3 *(Continued)*

Country	Agency	Software	Database
	Wuhan Technical University of Surveying and Mapping	Erdas Imagine, ER Mapper, Arc Info	
	National Remote Sensing Centre (Changsha)	Arc Info, City Star, MapGIS, LREIS Inttreal GIS, Mapcad	dBase, Foxpro
Indonesia	National Aeronautics and Space Institute	Metdas, Erdas, Dundee VISSR	
	Meteorological and Geophysical Agency		IQS, Oracle
Islamic Republic of Iran	Islamic Republic of Iran Meteorological Organization	UKW Technik, Bilko 1/2	
Japan	Japan Meteorological Agency		AIM/RDB
Malaysia	Malaysian Meteorological Service	Software developed by MacDonald Dettwiler and by Array Systems Computing	
Pakistan	Space and Upper Atmosphere Research Commission	Alden colour weather graphics processing software, Real Pak TIROS, TOVS and GEMPAK software, PC GEOS/WEFAX	
Philippines	Philippine Atmospheric, Geophysical and Astronomical Services Administration	NEC Basic, McIdas, GMS and NOAA processing systems	
Sri Lanka	Department of Meteorology	PCSAT	

Table 5.4 Sources and application of meteorological satellite data

Country	Agency	Data sources			Application		
		NOAA	GMS	Others	Wea-ther	Avi-ation	Others
Australia	Bureau of Meteorology	✓	✓		✓		
Azerbaijan	Azerbaijan National Space Agency	✓	✗	Meteor	✗	✓	Agriculture, defence, shipping
Bangladesh	Space Research and Remote Sensing Organization	✓	✓		✓	✗	Agriculture, disaster management
	National Satellite Meteorology Centre	✓	✓	FY-1	✓	✗	
	Institute of Remote Sensing Applications	✓	✗	GEO	✗	✗	Disaster management
China	Wuhan Technical University of Surveying and Mapping	✓	✓	FY-1	✗	✗	Mapping
India	Indian Space Research Organization	✓	✗	INSAT	✓	✓	Disaster management, shipping

211

Table 5.4 *(Continued)*

Country	Agency	Data sources			Application		
		NOAA	GMS	Others	Wea-ther	Avi-ation	Others
Indonesia	National Aeronautics and Space Institute	✓	✓		✗	✗	Agriculture
	Meteorological and Geophysical Agency	✓	✓	GRID	✗	✓	Agriculture, disaster management
Islamic Republic of Iran	Islamic Republic of Iran Meteorological Organization	✓	✓	Meteosat	✗	✓	Agriculture, media
Japan	Japan Meteorological Agency	✗	✓		✓	✗	
Malaysia	Malaysian Meteoro-logical Service	✓	✓		✓	✓	
Mongolia	National Remote Sensing Centre	✓	✓	Meteor	✓	✗	Agriculture, defence
Pakistan	Space and Upper Atmosphere Research Commission	✓	✗	ARGOS	✓	✓	Environment, shipping
Philippines	Philippine Atmospheric, Geophysical and Astro-nomical Services Administration	✓	✓		✓		
Republic of Korea	Korean Meteorological Research Institute	✓	✓		✗	✗	
Singapore	Meteorological Service Singapore	✓	✓		✗	✗	
Sri Lanka	Department of Meteorology	✓	✓	GOMS, INSAT	✗	✓	Fishing, shipping
Viet Nam	National Centre for Hydrometeorological Forecasting	✓	✓		✓	✗	Disaster management

Table 5.5 Estimated investment (US dollars) in satellite meteorology facilities

Country	Agency	In past 5 years	In next 5 years
Azerbaijan	Azerbaijan National Space Agency	1,000,000	1,500,000
Bangladesh	Space Research and Remote Sensing Organization	150,000	200,000
China	Institute of Remote Sensing Applications	1,200,000	1,200,000
	Wuhan Technical University of Surveying and Mapping	1,100,000	700,000
	National Remote Sensing Centre (Changsha)	95,000	120,000
Indonesia	National Aeronautics and Space Institute	2,000,000	4,000,000
Islamic Republic of Iran	Islamic Republic of Iran Meteorological Organization	800,000	0
Malaysia	Malaysian Meteorological Service	779,500	340,000
Pakistan	Space and Upper Atmosphere Research Commission	3,500,000	6,000,000
Philippines	Philippine Atmospheric, Geophysical and Astronomical Services Administration	Grant	500,000
Sri Lanka	Department of Meteorology	200,000	150,000
Viet Nam	National Centre for Hydrometeorological Forecasting	1,000,000	80,000

5.2 Country inputs

5.2.1 Australia

Bureau of Meteorology

The mandate of the Bureau of Meteorology is based on the Meteorology Act and subsequent agreements and decisions including the government's 1988 reaffirmation of the essentially public interest nature of meteorological data collection, research and services. It responds to Australia's ongoing national needs and international obligations which influences the goals, objectives and programmes of the Bureau. The Director of Meteorology is responsible for coordination of Australian involvement in the programmes and actvities of WMO.

Through its responsibilities under the Meteorology Act, the Bureau is concerned with, inter alia, the state of atmosphere, the interaction of the atmosphere with the underlying surface, the impact of weather and climate on human activity and the impact of human activities on the atmosphere. The pervasiveness of meteorological influences means that the Bureau's planning process must take account of the strategic and operational plans of many other bodies, including WMO.

Figure 5.1 shows the organizational chart of the Ministry of Environment, Sport and Territories. The programme structure and organization of the ministry at 30 June 1996 show the Bureau of Meteorology as both a statutory body reporting to the Parliamentary Secretary and a part of the ministry. More details may be found at the Bureau's Web site (http://www.bom.gov.au).

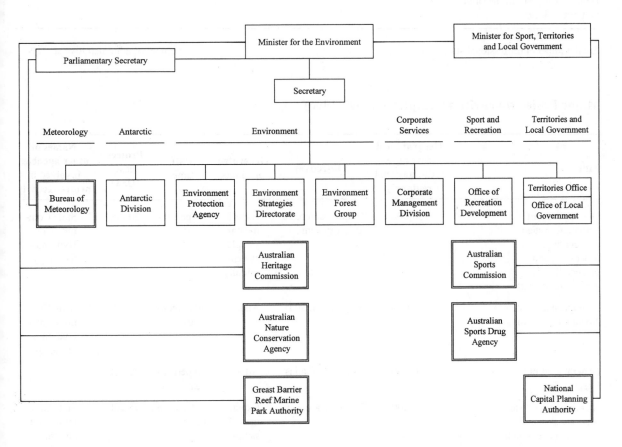

Figure 5.1 Organization chart of the Ministry of Environment, Sport and Territories

5.2.2 Azerbaijan

Azerbaijan National Space Agency

The Azerbaijan National Space Agency is a government organization whose mandate is specified in a decree issued by the President of Azerbaijan.

Professional Personnel

Primary field of work	Number of staff with the following qualifications			Number of staff with intensive training (>2 consecutive months) during last 5 years
	Ph.D.	MSc	BSc	
Weather forecasting				
Disaster monitoring/warning		2	10	
Environmental monitoring	1	24	60	
Climate change studies	1	2	12	
Hardware/software development	2	10	50	
Development of ground receiving and analysis stations		10	25	
Sensor development	2	20	100	
Radio communications				
Data management/information systems services				

Total number of professional staff: 331   Total number of administrative staff: N/A

Major Projects/Activities Completed (1992-1996)

Title	Geographical coverage (local, national, regional)	Objectives	Generated products	Staff time	Project cost (US$)	Names of other agencies involved in project/activity
Development of automatic system of receiving, processing and presentation of meteosatellite data	100,000	To develop/create such a system	Experimental sample system initially	60 person-years	20,000	State Committee on Hydrometeorology
Ecological atlas of Abaheron peninsula	5,000	To create an atlas	Atlas	60 person-years	80,000	State Committee on Hydrometeorology
General atlas of Caspian Sea	200,000	To create an atlas	Atlas	60 person-years	100,000	
Creation of hardware/software system for collection and digital processing of agro-met data of underlying surfaces	50,000	To create a system	Experimental sample system initially	116 person-years	45,000	State Committees on Hydrometeorology

Seminars/Conferences/Workshops Organized (1992-1996)

Type (seminar, conference, workshop, etc.)	Theme/title	Co-sponsorship (if any)	Number of participants	Title of output (publication and/or other)
Conference	Modern ecology problems	State Committee on Ecology	124	*Transactions,* 1994

Training Activities (1992-1996)

Course title and description	Training provided by organization for external users		
	Name of participating agencies/organizations	Duration/ frequency	Number of participants
Earth study from space; aerospace device development; measure-informative systems; automatic systems for information processing and control	Azerbaijan State Oil Academy	Annual	50

Databases/Archives Developed and/or Maintained and Available to Outside Users

Type (digital or analogue) and storage medium	Database contents (type, location, source data, scale and other useful information)	Geographical extent/surface area in km²	Year of production/ review	Internet address (URL)	Main users
Analogue	Data on temperature, thermo- and mass-exchange, wind speed, waving, Bowen numbers for Azerbaijan sector of Caspian Sea; format A4, 300 maps	20,000	1995		State Oil Company of Azerbaijan
Analogue	General atlas of Abaheron peninsula. Data on distribution of heavy metals, oil polluted regions, ecological state of plants, industrial zones, Baku and Sungait, format A4, scale 1:25,100 maps	50,000	1996		

Documents Published since 1990

Title	Year of publication	Language of publication	Keywords	Frequency
Transactions of Ecology Institute	1993	Russian	Oil pollution of land and water, heavy metals, land cover and degradation	
Transactions of Ecology Institute	1996	Russian		

5.2.3 Bangladesh

Bangladesh Space Research and Remote Sensing Organization

The Bangladesh Space Research and Remote Sensing Organization (SPARRSO) is an autonomous government organization mandated to apply space technology to surveying the natural resources and monitoring the environment and natural disasters in the country; to establish data acquisition, processing and dissemination system (satellite ground station), develop instrumentation facilities and trained manpower; to act as the national focal point for space and remote sensing activities in the country and to provide the government with relevant information in formulating national, regional and international policy issues concerning applications towards sustainable development and to establish regional and international cooperation and collaboration in the peaceful uses of space science and technology.

Professional Personnel

Primary field of work	Number of staff with the following qualifications			Number of staff with intensive training (>2 consecutive months) during last 5 years
	Ph.D.	MSc	BSc	
Weather forecasting	1	3	5	
Disaster monitoring/warning	1	2	5	
Environmental monitoring	2	5		
Climate change studies	1	1		
Hardware/software development		4		
Development of ground receiving and analysis stations				
Sensor development				
Radio communications				
Data management/information systems services		2	1	

Total number of professional staff: 33   Total number of administrative staff: 8

Major Projects/Activities Completed (1992-1996)

Title	Geographical coverage (local, national, regional)	Objectives	Generated products	Staff time	Project cost (US$)	Names of other agencies involved in project/activity
Upgrading of GMS ground station to receive GMS-5 data	Area covered by GMS-5 satellite	To use high and low resolution GMS-5 data in weather/cyclone foreasting and environmental monitoring	Imagery	3 person-years	Donated by United States Government	NASA

Major Current/Ongoing Projects/Activities

Title	Geographical extent/surface area in km²	Objectives	Generated products	Staff time	Project cost (US$)	Names of other agencies involved in project/activity
Monitoring of weather, especially the monsoons and cyclones	Bangladesh, Bay of Bengal and Far East	To assist in weather fore-casting and in cyclone and flood warning	Imagery			Bangladesh Meteorological Department, Water Board, Disaster Management Bureau
Study of marine environment of the Bay of Bengal using ADEOS OCTS data	5°-28°N 80°-105°E	To develop algorithms for SST, chlorophyll and turbidity in Bay of Bengal	SST, chlorophyll and turbidity algorithms and maps		40,000	ESCAP, NASDA

Seminars/Conferences/Workshops Organized (1992-1996)

Type (seminar, conference, workshop, etc.)	Theme/title	Co-sponsorship (if any)	Number of participants	Title of output (publication and/or other)
International seminar	Climate change and cyclones	Academy of Sciences	42	Proceedings to be published
Seminar	Monsoon dynamics	Humbolt Foundation	60	
International symposium	Natural disasters and their mitigation	Academy of Sciences, FASAS, FAO	50	

Documents Published since 1990

Title	Year of publication	Language of publication	Keywords	Frequency
SPARRSO Newsletter		English		Quarterly
Annual Report		English		Annual
Journal of Remote Sensing and Environment		English		Annual
Reports on various projects		English/ Bengali		Upon completion of project

Training Activities (1992-1996)

Course title and description	Training undergone by organization staff		
	Given by (name of institution/agency)	Duration/ frequency	Number of staff taking course
Training (associateship)	ICTP, Trieste, Italy	3 months	1

Course title and description	Training provided by organization for external users		
	Name of participating agencies/organizations	Duration/ frequency	Number of participants
Remote sensing and meteorology	Students of Dhaka University	6 months	4
Remote sensing and meteorology	Students of Khulna University	2 weeks	14
Remote sensing and meteorology	Students of Rajshahi Univesity	2 weeks	2

Databases/Archives Developed and/or Maintained and Available to Outside Users

Type (digital or analogue) and storage medium	Database contents (type, location, source data, scale and other useful information)	Geographical extent/surface area in km^2	Year of production/ review	Internet address (URL)	Main users
Digital	GMS data	Bangladesh and surrounding regions	Since 1991		SPARRSO, Bangladesh Meteorological Department, Agriculture and Food Ministry
Analogue	NOAA APT and AVHRR	Bangladesh and surrounding regions	Since 1978		SPARRSO, Bangladesh Meteorological Department, Agriculture and Food Ministry
Digital	NOAA AVHRR	Bangladesh and surrounding regions	Since 1985		SPARRSO, Bangladesh Meteorological Department, Agriculture and Food Ministry

5.2.4 China

National Satellite Meteorological Centre of China

The National Satellite Meteorological Centre of China is a national-level facility responsible for providing guidance in satellite meteorology to users in the whole of China.

Professional Personnel

Primary field of work	Number of staff with the following qualifications			Number of staff with intensive training (>2 consecutive months) during last 5 years
	Ph.D.	MSc	BSc	
Weather forecasting	2	7	23	
Disaster monitoring/warning		2	12	
Environmental monitoring		1	8	
Climate change studies		3	8	
Hardware/software development		2	25	
Development of ground receiving and analysis stations		4	14	
Sensor development				
Radio communications			8	
Data management/information systems services			10	

Total number of professional staff: 108   Total number of administrative staff: N/A

Major Projects/Activities (1992-1996)

Title	Geographical extent/surface area in km²	Objectives	Generated products	Staff time	Project cost (US$)	Names of other agencies involved in project/activity
FY-2 geostationary meteorological data reception and processing system			Digital products on clouds, winds, image graphics			
Upgrading of FY-1 satellite ground station						
Research programme on typhoons, rainstorms and other disasters of meteorological origin						
Sino-Japanese Asian monsoon research programme						

Seminars/Conferences/Workshops Organized (1992-1996)

Type (seminar, conference, workshop, etc.)	Theme/title	Co-sponsorship (if any)	Number of participants	Title of output (publication and/or other)
Seminar	Application of meteorological remote sensing on agriculture		40	Proceedings
Seminar	Strategy on meteorological satellite data processing and applications		35	

Training Activities (1992-1996)

Course title and description	Training provided by organization for external users		
	Name of participating agencies/organizations	Duration/ frequency	Number of participants
Training course on meteorological satellite data analysis and applications			72
Training course on meteorological satellite data for weather analysis and forecasting			53
Training course on meteorological Satellite data applications			60

Databases/Archives Developed and/or Maintained and Available to Outside Users

Type (digital or analogue) and storage medium	Database contents (type, location, source data, scale and other useful information)	Geographical extent/surface area in km²	Year of production/ review	Internet address (URL)	Main users
Digital Tape (6250)	NOAA-1B	China	1986-1988		Meteorological Administration; Agricultural Ministry; Forestry Ministry
Digital Tape (3480)	NOAA-1B, FY-1A, FY-1B, GMS-4, GMS-5	China	1988-1996		Meteorological Administration; Agricultural Ministry; Forestry Ministry

Documents Published since 1990

Title	Year of publication	Language of publication	Keywords	Frequency
FY-1 meteorological satellite data receiving and processing system	1990	Chinese		
Atlas of FY-1 meteorological satellite	1990	Chinese/English		
Atlas of meteorological satellites	1991	Chinese		
Technical report on satellite meteorology	1991-1996	Chinese		

Institute of Remote Sensing Applications

The Institute of Remote Sensing Applications is the leading government organization in the field of remote sensing applications, including the application of meteorological satellite data to weather and disaster monitoring. It also collaborates with other national and international organizations, including organizations in over 30 countries.

Professional Personnel

Primary field of work	Number of staff with the following qualifications			Number of staff with intensive training (>2 consecutive months) during last 5 years
	Ph.D.	MSc	BSc	
Weather forecasting				
Disaster monitoring/warning	6	8	2	
Environmental monitoring	7	8	2	
Climate change studies	5	3	2	
Hardware/software development	4	6	6	
Development of ground receiving and analysis stations	1	2	3	
Sensor development				
Radio communications				
Data management/information systems services				

Total number of professional staff: 125

Total number of administrative staff: 26

Major Projects/Activities Completed (1992-1996)

Title	Geographical coverage (local, national, regional)	Objectives	Generated products	Staff time	Project cost (US$)	Names of other agencies involved in project/activity
Drought monitoring and estimation through remote sensing	Regional	To develop an integrated drought monitoring and estimation system				SSTC, MAC, MFC
Technical system for multi-faceted remote sensing data acquisition	National					
GIS for disaster and yield estimation	National	To develop a GIS for monitoring and estimating disaster and yield				

Major Current/Ongoing Projects/Activities

Title	Geographical coverage (local, national, regional)	Objectives	Generated products	Staff time	Project cost (US$)	Names of other agencies involved in project/activity
Remote sensing and GIS for national basic farmland protection	National	To protect farmland	Software			
Earthquake prediction using thermal infra-red data	National	To predict earthquakes				

Seminars/Conferences/Workshops Organized (1992-1996)

Type (seminar, conference, workshop, etc.)	Theme/title	Co-sponsorship (if any)	Number of participants	Title of output (publication and/or other)
International conference	Microwave remote sensing		300 x 4 times	Proceedings
International workshop	Multi-angular remote sensing	SSTC	120	Proceedings

Training Activities (1992-1996)

Course title and description	Training undergone by organization staff		
	Given by (name of institution/agency)	Duration/ frequency	Number of staff taking course
Remote sensing applications	Canada Remote Sensing Centre	1 year	1
Geological remote sensing	ITC, Netherlands	4 years	3

Course title and description	Training provided by organization for external users		
	Name of participating agencies/organizations	Duration/ frequency	Number of participants
MSc. course			
Ph.D. course			

Databases/Archives Developed and/or Maintained and Available to Outside Users

Type (digital or analogue) and storage medium	Database contents (type, location, source data, scale and other useful information)	Geographical extent/surface area in km²	Year of production/ review	Internet address (URL)	Main users
Digital tape	GEO, AVHRR				
Digital tape	Airborne remote sensing data				
Digital	Database of China (1:250,000)				

Documents Published since 1990

Title	Year of publication	Language of publication	Keywords	Frequency
Journal of Remote Sensing	1990	Chinese (abstracts in English)		4 issues per year

5.2.5 India

SATELLITE METEOROLOGY/DISASTER MONITORING ACTIVITIES*

Introduction

Indian meteorologists have been using meteorological satellite data operationally since 1965. However, satellite meteorology in India may truly be said to have come of age in 1982 with the advent of India's own multi-purpose geostationary satellite system, Indian National Satellite (INSAT), which has radiometric imaging and data relay capabilities. Today, satellite observations have become an integral part of medium-range weather forecasting in the country. Presently India's INSAT and the United States' NOAA series are providing the basic observational data for operational meteorological forecasting in the country.

The India Meteorological Department is primarily responsible for meteorological forecasting, while the Department of Space provides satellite observation capabilities and carries out R and D work in meteorological sensors. The Department of Space also undertakes basic R and D activities in such areas as long-range weather prediction, modelling and other associated areas.

A. INSAT system

INSAT is a multi-purpose satellite system which has been providing telecommunications, broadcasting and meteorological services since 1983. While the first generation of satellites (INSAT-1) were procured from abroad, INSAT-2 satellites with higher capabilities have been developed indigenously. The present INSAT segment consists of one first-generation satellite, INSAT-1D, and four Indian-built second-generation satellites, INSAT-2A, 2B, 2C and 2D.

INSAT-1D, launched on 12 June 1990, is the last of the first-generation satellites still in service. It has the following radiometric imaging and data relay capabilities for meteorological applications, in addition to the services it is providing in telecommunications and television:

(a) A very high resolution radiometer (VHRR) for meteorological Earth imaging in visible (0.55-0.75 microns) and infra-red (10.5-12.5 microns) bands, with resolutions of 2.75 and 11 km respectively, and providing half-hourly full Earth coverage and sector scan capability;

(b) A data relay transponder having global receive coverage with a 402.75 MHz Earth to satellite link for relay of meteorological, hydrological and oceanographic data from unattended land- and ocean-based automatic data collection-cum-transmission platforms;

(c) A 406 MHz C-band Search and Rescue distress alert transponder.

INSAT-2A and 2B satellites, launched in 1992 and 1993, also have the above payloads with improved spatial resolutions of 2 km in the visible and 8 km in the near-infra-red bands. INSAT-2C and 2D, which are identical, were launched on 7 December 1995 and 4 June 1997. They do not have a meteorological payload since sufficient redundancy already exists. However, INSAT-2E, the last of the INSAT-2 series, scheduled for launch in 1998, will carry an improved meteorological payload which will include a water vapour channel (5.7 to 7.1 microns).

1. Meteorology services of INSAT

The meteorological component of the INSAT system provides:

(a) Round-the-clock, regular, half-hourly synoptic images of weather systems, including severe weather, cyclones, sea-surface and cloud-top temperatures, water bodies, snow, etc., over the entire territory of India as well as adjoining land and sea areas;

---

*Based on a report provided by the Indian Space Research Organization.

(b)   Collection and transmission of meteorological, hydrological and oceanographic data from unattended remote platforms;

(c)   Timely warnings of impending natural disasters like cyclones, storms and floods;

(d)   Dissemination of meteorological information, including processed images of weather systems to forecasting offices.

2. INSAT applications

INSAT images are of great value in observing the development of weather phenomena over oceanic areas and inaccessible regions where other observations are not available. Many different cloud systems have a distinct appearance in satellite pictures which makes it possible to identify jet streams, fog, thunderstorms, equatorial convergence zones, snowfall, etc. INSAT pictures are of particular importance in monitoring western disturbances, monsoon depressions, tropical cyclones and other migratory systems over land and sea. The meteorological data provided by INSAT benefits agriculture, aviation, ports and shipping, hydro-meteorological and flood forecasting services/sectors. With satellites it is possible to identify cyclones about 12-24 hours in advance and consequently issue early warnings.

3. Meteorological imagery

The VHRR instrument on board the INSATs provide half-hourly full-frame cloud imagery in visible and IR channels. The resolutions of the two channels at sub-satellite point are 2.75 and 11 km for INSAT-1, and 2 km and 8 km for INSAT-2 satellites. A limited sector scan of 6 minutes can also be taken. The primary data are subjected to detailed analysis using digital image processing techniques to provide the following main data products used in weather analysis and forecasting:

(a)   Cloud cover images in the visible and infra-red bands;
(b)   Cloud motion vectors (CMVs) giving winds in upper atmosphere;
(c)   Sea surface temperature;
(d)   Estimation of precipitation and outgoing long-wave radiation.

The cloud cover pictures are generated at least once every three hours daily. More frequent pictures are taken whenever the weather situation so demands. They are produced in visible and infra-red during the day and in infra-red at night. The images are used in conjunction with interactive facilities available at INSAT Meteorological Data Processing System (IMDPS) for the interpretation of cloud features. These images are particularly useful in identifying cloud systems over the ocean where no other observational data are available, for cyclone tracking, intensity assessment and prediction of storm surges, etc. CMVs are derived from consecutive pictures by observing movement of cloud clusters/tracers. Winds data derived at 0600 Z are being transmitted over GTS of WMO for operational use. The 0300 Z winds data are basically meant for national use. The CMVs are being derived by using a fully automated computer-based pattern-matching technique.

INSAT infra-red data are being used for derivation of sea surface temperature for the Bay of Bengal, the Arabian Sea and part of the Indian Ocean on an experimental basis. These data are a powerful tool for possible quantitative estimation of clouds. It has been possible to estimate the aerial precipitation produced by clouds using three-hourly INSAT infra-red data. Such precipitation estimates are accurate when they are averaged over large areas (2.5 degree square latitude/longitude) over 5 to 10 days or more. The precipitation estimates are required particularly over oceanic regions, where conventional precipitation measurements are not available because of problems connected with atmospheric modelling. Another product of interest for the atmospheric modelling and numerical weather prediction is the outgoing long-wave radiation (OLR). This is being derived over a 2.5 degree square latitude/longitude. Such data on precipitation and OLR provide valuable input to the numerical weather prediction model when used for issuing medium range weather forecasts (3-10 days ahead) for agricultural purposes and for climate modelling.

4. Data relay system

The data relay transponder (DRT) on board INSAT satellites is being used for collection of meteorological, hydrological and oceanographic data from remote, uninhabited locations over the land where about 100 data collection platforms are in operation.

5. Meteorological data dissemination service

The secondary data utilization centres at various forecasting centres of the India Meteorological Department receive processed cloud pictures, facsimile charts and conventional meteorological data transmitted from IMDPS directly from satellite using the S-band transponder of INSAT. Other institutions interested in meteorological information such as the agro-meteorological departments of agricultural universities also receive this signal and obtain meteorological information in near real time. A meteorological data dissemination terminal has been set up by India at Male, Maldives.

6. Cyclone warning dissemination system

The INSAT cyclone warning dissemination system (CWDS) makes use of the direct-to-community broadcast capability of the INSAT system. The system enables the Cyclone Warning Centre (CWC) to directly and selectively address a particular area likely to be hit by a cyclone. Simple receivers, which are an adaption of direct satellite community television receivers, are located in coastal villages. These receivers, designed for continuous operation in rural coastal environments are tuned to receive specific codes which are assigned to particular locations. CWC, after determining the likelihood of a cyclone hitting a place, selects the appropriate code and sends this signal to the satellite through an Earth station on an RN-like carrier. The satellite relays the signal back to the ground to be received by all receivers. But only those receivers tuned to the particular code transmitted activate a siren loud enough to be heard by people in the neighbourhood. The siren lasts about a minute and automatically switches off. CWC comes on the air after this with a warning in the local language giving details of the likely nature of the cyclone and the precautions to be observed. CWC can repeat this procedure as often as desired and also change the warning message as events develop. If the cyclone deviates from the course, CWC can direct the warning to another area and in fact gives out different warnings to different locations.

Two hundred and fifty DWS receivers are already in operation along the coastal belts of Tamil Nadu, Andhra Pradesh, Orissa, West Bengal, Gujarat and Karnataka. The INSAT CWDS, in conjunction with the VHRR imagery, was successfully used in issuing timely cyclone warnings to affected areas during several cyclones that have struck the eastern coastal regions of the country in the last ten years. CWDS has been instrumental in saving thousands of lives.

7. Satellite-aided search and rescue

India is a member of the international COSPAS-SARSAT programme which provides distress alert and position location service through low Earth orbit satellites. Appropriate ground segment has been operational since 1989.

While LEO satellites are well suited for this service owing to their ability to provide alert location information, the time gap between passes, particularly in the equatorial region, has been found quite high at times and the need for reducing such time gaps had been strongly felt. As the 406 MHz beacons, designed for the COSPAS-SARSAT system, operate at a fairly high power level (5W), it was recognized that using a 406 MHz receiver on a geostationary platform could help in almost-real-time detection of the distress signal, which could help in early mobilization of search and rescue efforts. INSAT-2A and 2B carry SAR payloads which receive emissions from the 406 MHz beacons, translate the signals to C-band and transmit the same to the ground station. The necessary ground station equipment (receiver and processor) are located at INSAT master control facility to detect the signals, which are then fed to the Indian COSPAS-SARSAT mission control centre at Bangalore.

B. NOAA applications

NOAA satellite data have been extensively used in monsoon studies, especially using NOAA AVHRR and TOVS data. The main product available from AVHRR used in monsoon studies is the sea surface temperature (SST). The channels used to estimate SST are centred around 3.71, 10.8, and 11.9 microns. SST is determined using a differential absorption technique to allow for atmospheric effects. The NOAA-SST is now available with an accuracy of 0.7°C in mid-latitudes and 1-2°C in the tropics.

TOVS-derived products like temperature and humidity profiles are used for understanding the monsoon circulation. The temperature profile mainly helps in examining the thermal features over the thermally strategic synoptic features like the heat low over deserts or the Tibetan high over high mountains. The humidity profile is useful in understanding the mechanism of moisture advection from the Indian Ocean to the subcontinent. TOVS-derived temperatures are available as mean layer temperatures ranging from surface to 1 mb (fifteen layers). The mean layer temperatures are generally believed to be accurate to about 1-2°K. The humidity is available only over three layers, i.e. surface to 700 mb, 700-500 mb, 500-300 mb layers.

The most important and probably the most significant contribution of satellites in meteorology has been the monitoring and tracking of tropical cyclones through satellite images since the launch of TIROS-1 in 1960. It is only since the advent of satellites that accurate global monitoring of tropical cyclones has been possible. Since tropical cyclones form over oceans, where conventional surface observations are scarce, meteorological satellites have provided a variety of information on cyclone intensity, movement, structure, etc., with high repetitiveness. A large number of synoptic case studies on cyclones using satellite images, radar and other conventional observations have been made all over the world and also in India. With the availability of data from a variety of satellite sensors (in the visible, infra-red and microwave regions of the electromagnetic spectrum – and active and passive sensors) it has become possible in recent years to obtain quantitative information of various cyclone parameters. Even from a set of two operational satellites, viz NOAA (polar orbiting) and INSAT (geostationary) a variety of observations on cyclones can be made.

C. Future plans during the next five years

(a) Efforts during the next five years will be directed towards providing data from INSAT and IRS satellites for the following thrust areas of applications in meteorology:

- Refinement of existing models and development of new assimilation techniques for improved weather forecasting

- Forecasting of tropical disturbances and cyclones in the medium range (development of techniques for synthetic generation of cyclone vortex based on a few key parameters such as its location, maximum wind speed, radius of maximum wind, centre pressure for accurate cyclone track prediction)

- Extended range monsoon prediction model for selected regions in the country; development of a coupled ocean-atmosphere-land model to predict seasonal/ monthly rainfall over homogeneous regions of India during the south-west monsoon season

- Refinement of techniques for ocean parameter retrieval

- Forecasting of ocean state variables/ocean weather services

(b) Snowcasting (from present time to a couple of hours);

(c) Short-range weather forecasts (up to about 2 days in advance);

(d) Medium-range weather forecasts (3-7 days);

(e) Long-range weather forecasts (seasonal and decadal);

(f) Efforts are continuing to realize a few indigenous satellites, specifically for addressing various ocean and atmospheric parameters, which lead to better space based observation capabilities for meteorology;

(g) Oceansat-l with payloads like ocean colour monitor and multifrequency scanning microwave radiometer is planned for launch in 1998. Plans are also under way to launch Oceansat-2 with microwave payloads such as scatterometer, altimeter and passive microwave radiometer;

(h) Recognizing the need for an exclusive satellite for climate-related applications, the Climatsat Programme Definition Group has been constituted to define a priority list of payloads for inclusion in the indigenous satellite for Climatsat;

(i) The addition of specific meteorological payloads are being planned in the future INSAT series such as additional water vapour channel in VHRR payload and a charge coupled device (CCD) camera in INSAT-2E and atmospheric sounders in INSAT-3 series;

(j) These indigenous planned satellites, along with other international satellite missions, will provide improved observational capabilities, besides providing data on a regular basis.

D. Infrastructure and facilities

The India Meteorological Department is primarily responsible for meteorology and climate studies. This Department established a Meteorological Data Utilization Centre in 1982 at New Delhi to process the data from VHRR and DRT. This Centre has the facility to receive, process and analyse satellite data from INSAT and NOAA satellites. There are several regional centres with facilities for satellite data analysis for various applications. INSAT VHRR data are available in near real time at 25 meteorological data dissemination centres (MDDC) in various parts of the country. With the commissioning of direct satellite retransmission service of processed VHRR data over a CxS link, MDDC-type data can be provided at any location in the country.

1. Facilities at Department of Space centres

The following facilities are available at various Department of Space centres for carrying out R and D activities related to meteorology.

(a) *Space Applications Centre, Ahmedabad*

The Space Applications Centre and Ahmedabad has the facility to receive NOAA and INSAT data in real time and carry out various applications studies and related research. The thrust areas are tropical cyclone monitoring; rainfall estimation from satellite data; study of monsoon weather systems; and geosphere-biosphere studies.

The Meteorology and Oceanography Group at the Centre is also involved in analysis and assimilation of satellite data in numerical global circulation models for forecasting in extended range.

(b) *Space Physics Laboratory, Thiruvanathapuram*

Scientists at the Space Physics Laboratory at Thiruvanathapuram are engaged in the study of atmospheric processes and related energy budget. Tower-based measurements in conjunction with satellite data are used to study atmospheric dynamics. Aerosol estimation and atmospheric boundary layer are also studied.

(c) National Remote Sensing Agency

The Agency has the facility to receive and process NOAA data. The DPS system is used to generate vegetation maps to study drought conditions across the country. Sea surface temperature maps are generated on a weekly basis and used for various applications including fisheries potential assessment.

(d) SHAR rocket launching station

The meteorological facility is used to monitor weather conditions continuously, especially wind shear, over the launch station. Meteorological rockets are launched under a scientific programme to study troposphere and stratosphere conditions and processes.

E. United Nations-affiliated space technology and education centre

The Centre for Space Science and Technology Education in Asia and the Pacific (affiliated to the United Nations) was established in India at the initiative of the United Nations and through an intergovernmental agreement. It will offer education, research and applications experience to participants in all its programmes.

The Centre's campus is located in Dehra Dun around the infrastructure available at the Indian Institute of Remote Sensing. Further, the Centre has arrangements with Space Applications Centre at Ahmedabad, which acts as the host institution for courses and programmes related to satellite communications and satellite meteorology, and with the Physical Research Laboratory at Ahmedabad for courses and programmes in space sciences.

As the host country for the Centre, India has provided support and facilities, faculty, data, international hostel, fellowship etc., to the Centre for the conduct of educational programmes. The first course on remote sensing and GIS started from 1 April 1996 at IIRS. Twenty-five students from 14 countries attended the course and passed out with diplomas in December 1996. Facilities at IIRS were upgraded to cater to the requirements of the international course. The second education course on satellite communications was held at the Space Applications Centre, Ahmedabad, from January to September 1997. Fourteen students from nine countries attended the course. Facilities at the Centre were upgraded for the course. It is also proposed to conduct a nine-month training course on satellite meteorology and global climate from March 1998 at the Space Applications Centre under the aegis of the United Nations-affiliated Centre for Space Technology Training and Education (CSSTE-AP) for the benefit of the countries in the Asian and Pacific region.

5.2.6 Indonesia

National Institute of Aeronautics and Space

The National Institute of Aeronautics and Space (LAPAN) is a government organization mandated to undertake R and D in remote sensing technology and its application, and serve as the national remote sensing data bank.

Major Projects/Activities Completed (1992-1996)

Title	Geographical coverage (local, national, regional)	Objectives	Generated products	Staff time	Project cost (US$)	Names of other agencies involved in project/activity
Upgrading of NOAA and GMS stations	National	To improve the capability of the NOAA and GMS stations	Digital data (NOAA AVHRR, GMS VIS/IR)	10 person-years	1,000,000	MDA, Canada; University of Dundee, United Kingdom

Major Current/Ongoing Projects/Activities

Title	Geographical coverage (local, national, regional)	Objectives	Generated products	Staff time	Project cost (US$)	Names of other agencies involved in project/activity
Drought monitoring	Local	To improve capability to receive information about dry or drought-affected areas in Java and Sumatra in dry season	Regular reports	2.5 person-years	2,000	
Forest fire monitoring	National	To detect and monitor forest fires	Regular reports	2.5 person-years	2,000	
Monitoring of ITCZ and SPCZ movement	National	To predict the end of the rainy season	Regular reports	5 person-years	2,000	
Establishment of Earth observation centre	National	To establish such a centre, including provision of buildings, infrastructure, facilities, etc.		10 person-years	750,000	

Databases/Archives Developed and/or Maintained and Available to Outside Users

Type (digital or analogue) and storage medium	Database contents (type, location, source data, scale and other useful information)	Geographical extent/surface area in km^2	Year of production/ review	Internet address (URL)	Main users
Digital, CCT and CD	NOAA AVHRR		1993		Internal researchers/ scientists
Digital, magnetic optical disk	GMS VISSR	Global	1993		Internal researchers/ scientists

Meteorological and Geophysical Agency

The Meteorological and Geophysical Agency is a government organization mandated to provide national weather, climate and seismic services; promote data exchange with other national weather bureaux; and serve on the World Meteorological Organization.

Professional Personnel

Primary field of work	Number of staff with the following qualifications			Number of staff with intensive training (>2 consecutive months) during last 5 years
	Ph.D.	MSc	BSc	
Weather forecasting	1	2	3	2: numerical weather
Disaster monitoring/warning		1	4	
Environmental monitoring		2	2	
Climate change studies		2	2	
Hardware/software development		1		
Development of ground receiving and analysis stations				
Sensor development				
Radio communications				
Data management/information systems services		1	3	

Total number of professional staff: 142

Total number of administrative staff: 61

Major Projects/Activities Completed (1992-1996)

Title	Geographical coverage (local, national, regional)	Objectives	Generated products	Staff time	Project cost (US$)	Names of other agencies involved in project/activity
SEISCOM	National	To develop earthquake monitoring system		17 person-years	50,000,000	
SYNERGIC	National	To provide graphical integrated meteorology display system		13 person-years	10,000,000	

Major Current/Ongoing Projects/Activities

Title	Geographical coverage (local, national, regional)	Objectives	Generated products	Staff time	Project cost (US$)	Names of other agencies involved in project/activity
Meteorology data acquisition network	National	To collect meteorological data using VSAT	Meteo-rology/ rainfall database	13 person-years	20,000,000	

Seminars/Conferences/Workshops Organized (1992-1996)

Type (seminar, conference, workshop, etc.)	Theme/title	Co-sponsorship (if any)	Number of participants	Title of output (publication and/or other)
Workshop	International workshop on monsoon studies	WMO	150	Proceedings
Seminar	Meteorology and the media		50	
Press conference	Dry season		50	Publication

Training Activities (1992-1996)

Course title and description	Training undergone by organization staff		
	Given by (name of institution/agency)	Duration/ frequency	Number of staff taking course
NOAA ground receiving station (1995)	NRI/United Kingdom	2 weeks	10

Databases/Archives Developed and/or Maintained and Available to Outside Users

Type (digital or analogue) and storage medium	Database contents (type, location, source data, scale and other useful information)	Geographical extent/surface area in km^2	Year of production/ review	Internet address (URL)	Main users
Digital, magnetic tape	Rainfall, other data	Point/local	1990-1995	bmg.cbn. net.id	Public Works, Agriculture
Digital, DAT	Seismic data	Point/local	Current	bmg.cbn. net.id	Public Works

Documents Published since 1990

Title	Year of publication	Language of publication	Keywords	Frequency
Information on meteorology	Current	Indonesian	Rainfall	Monthly
Prediction of season	Current	Indonesian	Rainfall	Bi-annual
Weather reports	Current	Indonesian	Weather, wind, temperature	Daily
Buku Iklim	1990	Indonesian	Rainfall, temperature	Annual

5.2.7 Islamic Republic of Iran

Islamic Republic of Iran Meteorological Organization

The Islamic Republic of Iran Meteorological Organization (IRIMO) is a government organization mandated to perform analysis of atmosphere and related phenomena; organize information and data in support of agriculture, transportation, water, energy, environment protection, industry and related sectors; and promote data exchange and conduct research projects related to WMO programmes.

Professional Personnel

Primary field of work	Number of staff with the following qualifications			Number of staff with intensive training (>2 consecutive months) during last 5 years
	Ph.D.	MSc	BSc	
Weather forecasting		17	16	
Disaster monitoring/warning		17	16	
Environmental monitoring	3	9	19	
Climate change studies	3	10		
Hardware/software development			26	
Development of ground receiving and analysis stations			4	
Sensor development			4	
Radio communications			6	
Data management/information systems services		2	9	
Marine sciences	5	5		

Total number of professional staff: 1,034   Total number of administrative staff: 318

Seminars/Conferences/Workshops Organized (1992-1996)

Type (seminar, conference, workshop, etc.)	Theme/title	Co-sponsorship (if any)	Number of participants	Title of output (publication and/or other)
Conference	First regional conference on climate change	WMO, UNESCO	500	Proceedings or abstracts
Weekly seminars	Various topics of research			
Daily routine briefings				

Databases/Archives Developed and/or Maintained and Available to Outside Users

Type (digital or analogue) and storage medium	Database contents (type, location, source data, scale and other useful information)	Geographical extent/surface area in km^2	Year of production/ review	Internet address (URL)	Main users
Analogue, charts	Synpotic, Skew-T	70°W-90°E	1971-1997		IRIMO
Digital, magnetic tape	Synpotic climatic, rain gauge station data	Whole country	1951-1997 (every 5 years)		IRIMO
Analogue, hard copy	Meteosat VIS/IR data – 6 hourly	Middle East	1992 to present		IRIMO

Major Current/Ongoing Projects/Activities

Title	Geographical coverage (local, national, regional)	Objectives	Generated products	Staff time	Project cost (US$)	Names of other agencies involved in project/activity
Temporal variations of temperature	Local/regional	Research				
Anomalous warm zones	Local/regional	Research				
Wind tracking	Local/regional	Research				
Sea surface temperature measurements	Local/regional	Research				
Water vapour absorption patterns	Local/regional	Research				
Digital animation and image processing experiments	Local/regional	Research				
Dust storm monitoring	Local/regional	Research				
Atmospheric subsidence	Local/regional	Research				
Thermal inertia analysis	Local/regional	Research				
Marine research	Local/regional	Research				

Training Activities (1992-1996)

Course title and description	Training undergone by organization staff		
	Given by (name of institution/agency)	Duration/ frequency	Number of staff taking course
WMO Courses for IRIMO personnel at various levels	National experts in relevant fields		479
Ph.D. and MSc courses in atmospheric and marine sciences	Universities of Tehran and the Philippines		
Short-term courses in the Islamic Republic of Iran and abroad			300

I

Course title and description	Training provided by organization for external users		
	Name of participating agencies/organizations	Duration/ frequency	Number of participants
Satellite meteorology course	WMO nominees	2 weeks/year	25
CUCOM course		2 weeks/year	40
Agricultural meteorology course		2 weeks/year	20
Sandwich courses on weather and marine forecasting		2 months	70

Documents Published since 1990

Title	Year of publication	Language of publication	Keywords	Frequency
Weekly, monthly, seasonal and annual bulletins	1990 to present	Persian	Meteorology	
Reports of normal climatic standards of various stations		Persian	Meteorology	
NIVAAR (scientific journal)		Persian	Meteorological	

5.2.8 Japan

Japan Meteorological Agency

In the framework of the World Weather Watch programme promoted by the World Meteorological Organization, the Japan Meteorological Agency operates the Geostationary Meteorological Satellite (GMS) series.

Major Projects/Activities Completed (1992-1996)

Title	Geographical coverage (local, national, regional)	Objectives	Generated products	Staff time	Project cost (US$)	Names of other agencies involved in project/activity
Launch of GMS-5 and upgrading of ground facilities	National and regional	Weather monitoring	VISSR WEFAX SST, etc.		170,000	

Major Current/Ongoing Projects/Activities

Title	Geographical coverage (local, national, regional)	Objectives	Generated products	Staff time	Project cost (US$)	Names of other agencies involved in project/activity
Launch of multi-functional transport satellite and upgrading of ground facilities	National and regional	Weather monitoring	HiRID, LRIT, SST, etc.			

Seminars/Conferences/Workshops Organized (1992-1996)

Type (seminar, conference, workshop, etc.)	Theme/title	Co-sponsorship (if any)	Number of participants	Title of output (publication and/or other)
Seminar	Utilization of extended information from GMS	Ship and Ocean Foundation	13	Proceedings of the seminar

Training Activities (1992-1996)

| Course title and description | Training provided by organization for external users | | |
	Name of participating agencies/organizations	Duration/ frequency	Number of participants
JICA group training course in meteorology	National meteorological services/departments of 24 countries	Annual	39
JICA country-focused training course on weather satellites for the People's Republic of China	Satellite Meteorological Administration/Satellite Meteorological Centre		5
Asia-Pacific satellite applications training seminar	National meteorological services/departments of 20 countries		60

Documents Published since 1990

Title	Year of publication	Language of publication	Keywords	Frequency
Meteorological Satellite Centre Technical Notes	Since 1980	Japanese (abstracts in English)		Bi-annual

Databases/Archives Developed and/or Maintained and Available to Outside Users

Type (digital or analogue) and storage medium	Database contents (type, location, source data, scale and other useful information)	Geographical extent/surface area in km^2	Year of production/ review	Internet address (URL)	Main users
Digital, magnetic tape	VISSR image data (visible), MAX 50 MB/File	Full disk	1987		
Digital, magnetic tape	VISSR image data (IR), MAX 9.2 MB/File	Full disk	1981		
Digital, magnetic tape	VISSR image data (water vapour), MAX 9.2 MB/File	Full disk	1986		
Digital, magnetic tape	VISSR image data (typhoon special observation), MAX 67 MB/File	Northern hemisphere	1987		
Digital, magnetic tape	Cloud motion wind, MAX 52 MB/File	50°N-49°S 90°E-171°W	1987		
Digital, magnetic tape	Sea surface temperature, MAX 1.2 MB/File	50°N-50°S 90°E-170°W	1987		
Digital, magnetic tape	Cloud amount, MAX 9.2 MB/File	59.5°N-59.5°S 80.5°E-160.5°W	1987		
Digital, magnetic tape	Basic histogram, MAX 11.1 MB/File	Full disk	1978		
Digital, magnetic tape	Cloud information chart, MAX 0.1 MB/File	Around Japan	1986		
Digital, magnetic tape	Long wave radiation, MAX 3.3 MB/File	60°N-60°S 80°E-160°W	1987		

5.2.9 Malaysia

Malaysian Meteorological Service

The Malaysian Meteorological Service is a government organization mandated to provide satellite images to assist in the maintenance of a continuous weather watch over Malaysia; provide satellite information for aviation; and detect forest fires and open burning activities.

Professional Personnel

Primary field of work	Number of staff with the following qualifications			Number of staff with intensive training (>2 consecutive months) during last 5 years
	Ph.D.	MSc	BSc	
Weather forecasting				
Disaster monitoring/warning			1	
Environmental monitoring				
Climate change studies				
Hardware/software development				
Development of ground receiving and analysis stations			1	
Sensor development				
Radio communications				
Data management/information systems services		1		

Total number of professional staff: 3   Total number of administrative staff: 9

Major Current/Ongoing Projects/Activities

Title	Geographical coverage (local, national, regional)	Objectives	Generated products	Staff time	Project cost (US$)	Names of other agencies involved in project/activity
NDVI over MADA region	Subnational	To forecast yields through NDVI monitoring				Department of Agriculture
Forest fire monitoring	National	To forest fires and open burning activities				Department of Environment

Databases/Archives Developed and/or Maintained and Available to Outside Users

Type (digital or analogue) and storage medium	Database contents (type, location, source data, scale and other useful information)	Geographical extent/surface area in km^2	Year of production/ review	Internet address (URL)	Main users
Digital, CCT	Visible and infra-red data over the ASEAN region	ASEAN region	1990-1991		Internal
Digital, CCT	Data from NOAA AVHRR channels 1 to 5 of Malaysia	Whole of Malaysia	1994-1997		University

5.2.10 Mongolia

National Remote Sensing Centre

The National Remote Sensing Centre is a government organization mandated to coordinate activities in the country related to meteorology and disaster monitoring using satellite data; to develop a natural disaster monitoring system; to provide information on disasters.

Professional Personnel

Primary field of work	Number of staff with the following qualifications			Number of staff with intensive training (>2 consecutive months) during last 5 years
	Ph.D.	MSc	BSc	
Weather forecasting			1	
Disaster monitoring/warning	1	1	1	
Environmental monitoring	2	3	4	
Climate change studies				
Hardware/software development		1	1	
Development of ground receiving and analysis stations				
Sensor development				
Radio communications				
Data management/information systems services		1	1	

Total number of professional staff: 17

Total number of administrative staff: 2

Major Projects/Activities Completed (1992-1996)

Title	Geographical coverage (local, national, regional)	Objectives	Generated products	Staff time	Project cost (US$)	Names of other agencies involved in project/activity
Snow monitoring	National	To produce snow cover maps using AVHRR data		3 person-years	2,000	
Vegetation and grassland estimation	Regional	To study condition of grasslands and estimate biomass		3 person-years	3,000	
Rainfall prediction method	Local	To predict thunderstorms using satellite and radar data		2 person-years	1,500	Institute of Meteorology
Fire monitoring	National	To detect fires and assist in assessing environmental impact of fires		5 person-years	3,000	Civil Defence Office
Heavy snow monitoring	National	To produce snow classification maps		3 person-years	2,000	
Drought monitoring	National	To produce drought maps and assist in assessing effect of drought on agriculture and desertification		3 person-years	3,000	Institute of Agriculture

Major Current/Ongoing Projects/Activities

Title	Geographical coverage (local, national, regional)	Objectives	Generated products	Staff time	Project cost (US$)	Names of other agencies involved in project/activity
Natural disaster monitoring	National	To study the complex systems of natural disasters and develop disaster management systems		5 person-years	5,000	
Land assessment and classification	National	To produce land cost and classification maps		2 person-years	10,000	
Environmental database based on remote sensing and GIS, including natural disaster and meteorology database	National	To develop databases		8 person-years	15,000	About 21 organizations
Reforestation after fire	Local	To assist in the reforestation efforts after forest fires		2 person-years		Institute of forests

Seminars/Conferences/Workshops Organized (1992-1996)

Type (seminar, conference, workshop, etc.)	Theme/title	Co-sponsorship (if any)	Number of participants	Title of output (publication and/or other)
Workshop	Using NAS and satellite data receiving system	NASA	15	Users' guide
Workshop	Fire monitoring		35	
Training course	Database management system for meteorology and climatology		7	Users' guide
Seminar	Natural disaster monitoring			

Databases/Archives Developed and/or Maintained and Available to Outside Users

Type (digital or analogue) and storage medium	Database contents (type, location, source data, scale and other useful information)	Geographical extent/surface area in km²	Year of production/ review	Internet address (URL)	Main users
Analogue	NOAA AVHRR/HRPT image archive	35°-85°N 80°-130°E	1971-1988		Ministries of Nature and Environment, Agriculture and Transport
Digital	NOAA AVHRR/HRPT image tape	35°-85°N 80°-130°E	1988 onwards		Public
Digital	Remote sensing and GIS-based environmental database, including disaster, climatology and meteorology	Whole of Mongolia			Public

M

Training Activities (1992-1996)

Course title and description	Training undergone by organization staff		
	Given by (name of institution/agency)	Duration/ frequency	Number of staff taking course
Satellite data applications	JICA	2 months	1
Applications of TOVS data	China Meteorological Agency	2 weeks	1
Applications of satellite data in the Asian and Pacific region	WMO	2 weeks	1
Incident command system	Government of Mongolia	1 week	2

Course title and description	Training provided by organization for external users		
	Name of participating agencies/organizations	Duration/ frequency	Number of participants
Using NAS system	Institute of Meteorology	1 week	15
Fire monitoring	Civil Defence Office	1 week	35
Natural disaster reduction	Ministry of Nature and Environment and environmental organizations	3 days	20

Documents Published since 1990

Title	Year of publication	Language of publication	Keywords	Frequency
Various reports, papers relating to satellite meteorology and disaster monitoring		Mongolian, Russian and English		

5.2.11 Pakistan

Space and Upper Atmosphere Research Commission

The Space and Upper Atmosphere Research Commission (SUPARCO) is an autonomous commission under the Government of Pakistan. As the national space agency, SUPARCO is responsible for the implementation of the national space programme, which is essentially aimed at furthering research in space science and related fields, enhancing capabilities in space technology and promoting the applications of space science and technology for the socio-economic development of the country.

Professional Personnel

Primary field of work	Number of staff with the following qualifications			Number of staff with intensive training (>2 consecutive months) during last 5 years
	Ph.D.	MSc	BSc	
Weather forecasting			2	
Disaster monitoring/warning		2	4	
Environmental monitoring	1	2	2	
Climate change studies				
Hardware/software development		2	3	
Development of ground receiving and analysis stations				
Sensor development				
Radio communications				
Data management/information systems services		1	3	

Total number of professional staff: 22

Total number of administrative staff: 7

Major Projects/Activities Completed (1992-1996)

Title	Geographical coverage (local, national, regional)	Objectives	Generated products	Staff time	Project cost (US$)	Names of other agencies involved in project/activity
Summer monsoon monitoring over Pakistan	National	To identify patterns for accurate short- and long-range forecasting	APT images, TOVS charts and flow diagrams	3 person-years		
Total ozone monitoring over Karachi	Local	To examine seasonal and inter-annual variations of O_3 over Karachi	TOVS and sunspots data	3.5 person-years		
Identifying potential solar power sites	National	To identify sites suitable for solar power generation through study of cloud cover	APT images	0.5 person-year		

P

Title	Geographical coverage (local, national, regional)	Objectives	Generated products	Staff time	Project cost (US$)	Names of other agencies involved in project/activity
Pre-monsoon SST variability over EEZ of Pakistan		To study SST variability in pre-monsoon period	APT images and TOVS charts	2.5 person-years		
Air-sea interaction studies	National/regional			2 person-years		
Cylone monitoring	Regional	To track movement and make timely forecasts of cyclones		2 person-years		

Major Current/Ongoing Projects/Activities

Title	Geographical coverage (local, national, regional)	Objectives	Generated products	Staff time	Project cost (US$)	Names of other agencies involved in project/activity
Summer monsoon monitoring	Regional	To identify patterns for accurate short-, medium- and long-range forecasting	APT images, TOVS charts, flow diagrams and forecast bulletins	2 person-years		
Total ozone monitoring over Pakistan and adjoining areas	National/regional	To develop regression model for determining geographical and seasonal variations in total O_3 over Pakistan and adjoining areas	TOVS charts and flow diagrams	2.5 person-years		

Seminars/Conferences/Workshops Organized (1992-1996)

Type (seminar, conference, workshop, etc.)	Theme/title	Co-sponsorship (if any)	Number of participants	Title of output (publication and/or other)
Seminar	Seminar on space technology and its significance for Pakistan (Lahore, 1994)	Institute of Engineers, Pakistan	100	Proceedings published by SUPARCO
Conference	Second Asia-Pacific conference on multilateral cooperation in space technology and applications (Islamabad, 1995)	Space agencies in China, Republic of Korea and Thailand	110	Proceedings published by SUPARCO

Training Activities (1992-1996)

Course title and description	Training undergone by organization staff		
	Given by (name of institution/agency)	Duration/ frequency	Number of staff taking course
Atmospheric chemistry	Visiting Professor from State University of New York	1 month	10

Course title and description	Training provided by organization for external users		
	Name of participating agencies/organizations	Duration/ frequency	Number of participants
APT system, data analysis and interpretation	Various national organizations concerned with meteorology	2-3 weeks, twice a year	5-10 per course

Databases/Archives Developed and/or Maintained and Available to Outside Users

Type (digital or analogue) and storage medium	Database contents (type, location, source data, scale and other useful information)	Geographical extent/surface area in km^2	Year of production/ review	Internet address (URL)	Main users
Analogue in audio casettes and hardcopy in album form	APT (VIS and IR) data from NOAA series of satellites and from METEOR, acquired at Karachi. The APT station has been in operation round the clock since its installation in 1987.	3-4 million	1987 to date		Pakistan Meteorological Department, Pakistan International Airlines and various agencies concerned with aviation, meteorology and shipping
Digital in TK-70 tape cartridges and diskettes	TOVS (HIRS and MSU) 24 channel data from NOAA series of satellites. The TOVS station has been in operation since 1993.	2-3 million	1993 to date		Pakistan Meteorological Department, Pakistan International Airlines and various agencies concerned with aviation, meteorology and shipping

Documents Published since 1990

Title	Year of publication	Language of publication	Keywords	Frequency
Space Horizons, research journal	1984 onwards	English	Research papers, reports, articles, etc. on various aspects of space science, technology and applications	Quarterly
SUPARCO News (previously *SUPARCO Times*), newsletter on SUPARCO activities	1984 onwards	English	Activities, events, projects, developments, personnel, etc. concerning SUPARCO	Monthly

P

Title	Year of publication	Language of publication	Keywords	Frequency
Space Research in Pakistan, biennial national report submitted to the Committee on Space Research (COSPAR)	Since the 1960s	English	Pakistan's space programme: organization, establishments, R and D activities, projects, facilities, seminars/ conferences, training, international cooperation	Biennial
Proceedings of the Seminar on Space Technology and its Significance for Pakistan	1994	English	Space technology, communications, remote sensing for resource/environmental surveying, satellite meteorology	
Proceedings of the Second Asia-Pacific Conference on Multilateral Cooperation in Space Technology and Applications	1996	English	Space technology, its applications, Asia-Pacific regional cooperation, institutional framework, small multi-mission satellite project	
ISNET News, newsletter covering activities of Inter-Islamic Network on Space Sciences and Technology coordinated by Pakistan		English	News about activities, events, developments relating to ISNET	Bi-annual
Various research papers, status reports, country reports, opinion papers, etc. published in national and international journals and proceedings of seminars/ conferences		English		

5.2.12 Philippines

Philippine Atmospheric, Geophysical and Astronomical Services Administration

The Philippine Atmospheric, Geophysical and Astronomical Services Administration (PAGASA) is a government organization mandated to undertake activities related to observation, collection, assessment and processing of atmospheric and allied data for the benefit of agriculture, commerce and industry; engage in studies of atmospheric, geophysical and astronomical phenomena essential to the safety and welfare of the people as well as undertake researches on the structure, development and motion of typhoons and formulate measures for their mitigation; maintain a nationwide network of monitoring stations in support of forecasting weather, flood and other climatological conditions affecting the national safety and economy; and maintain effective linkages with scientific organizations and promote exchange of scientific information and cooperation among personnel engaged in atmospheric, geophysical, and astronomical and space studies. Its organization chart is shown in figure 5.2.

Professional Personnel

Primary field of work	Number of staff with the following qualifications			Number of staff with intensive training (>2 consecutive months) during last 5 years
	Ph.D.	MSc	BSc	
Weather forecasting	6	2	9	2
Disaster monitoring/warning				
Environmental monitoring				
Climate change studies				
Hardware/software development				
Development of ground receiving and analysis stations				
Sensor development				
Radio communications				
Data management/information systems services				

Total number of professional staff: 17   Total number of administrative staff: 61

Major Projects/Activities in the last 5 years (1996-2000)

Title	Geographical extent/surface area in km²	Objectives	Generated products	Staff time	Names of other agencies involved in project/activity
ERS-1 SAR data for flood monitoring in the Bicol region	Regional	To evaluate the capability of ERS-1 SAR in flood monitoring	Maps showing extent of flooding		NAMRIA

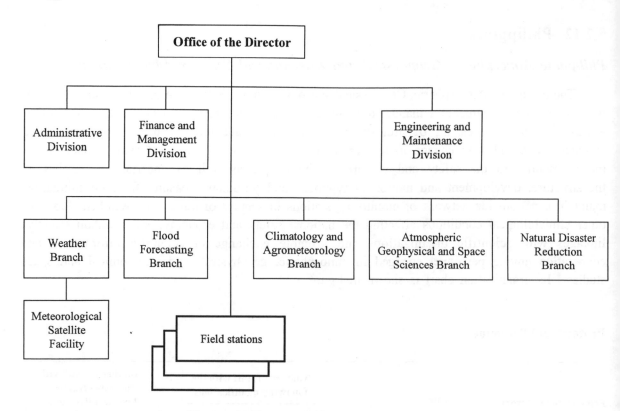

Figure 5.2 Organization chart of the Philippine Atmospheric, Geophysical and Astronomical Services Administration

Major Current/Ongoing Projects/Activities

Title	Geographical extent/surface area in km^2	Objectives	Generated products	Staff time	Names of other agencies involved in project/activity
Estimation of short-duration rainfall from GMS images		To estimate the short-duration rainfall (from hourly to daily) using GMS images			
Land-use/land-cover change using remote sensing and GIS		To determine land-use and land-cover change using remote sensing and GIS	Maps		NAMRIA, NEDA
Reception and processing of satellite images (visible, infra-red)	4,928,400	To provide hourly, 3-hourly (GMS) and 12-hourly (NOAA) satellite data for operational weather forecasting and research	Visible and IR images and TOVS data		

Training Activities (1992-1996)

Course title and description	Training undergone by organization staff		
	Given by (name of institution/agency)	Duration/ frequency	Number of staff taking course
Short course in remote sensing	University of the Philippines	2 weeks	2

Course title and description	Training provided by organization for external users		
	Name of participating agencies/organizations	Duration/frequency	Number of participants
Basic satellite image interpretation/operation and maintenance of WEFAX	National Power Corporation	2 weeks	4

Databases/Archives Developed and/or Maintained and Available to Outside Users

Type (digital or analogue) and storage medium	Database contents (type, location, source data, scale and other useful information)	Geographical extent/surface area in km^2	Year of production/review	Internet address (URL)	Main users
Digital – 8 mm Exabyte tape	Raw data (NOAA HRPT)		1992-1994		PAGASA, NAMRIA

P

5.2.13 Republic of Korea

Korea Meteorological Research Institute

The Korea Meteorological Research Institute is the national contact point for the regional working group on meteorological satellite applications and natural hazards monitoring.

Professional Personnel

Primary field of work	Number of staff with the following qualifications			Number of staff with intensive training (>2 consecutive months) during last 5 years
	Ph.D.	MSc	BSc	
Weather forecasting		4		
Disaster monitoring/warning		2		
Environmental monitoring	2			
Climate change studies	1	1		
Hardware/software development		1	5	
Development of ground receiving and analysis stations			5	
Sensor development				
Radio communications				
Data management/information systems services			5	

Total number of professional staff: 23

Total number of administrative staff: 2

Major Projects/Activities (1992-1996)

Title	Geographical coverage (local, national, regional)	Objectives	Generated products	Staff time	Project cost (US$)	Names of other agencies involved in project/activity
Development of very short-range precipitation forecasting system	Regional	To develop a precipitation forecasting system based on satellite and radar data		3 person-years		Seoul National University
Development of PC-based polar-orbiting meteorological satellite data analysis system	Regional	To develop a PC-based system for analysis of NOAA AVHRR data		1 person-year		
Estimation of precipitation intensity index using satellites	Regional	To estimate precipitation intensity index using satellite data		2 person-years		
Determination of typhoon centre using meteorological satellite data	Regional	To develop techniques for determining the position of typhoon centre		1 person-year		

Title	Geographical coverage (local, national, regional)	Objectives	Generated products	Staff time	Project cost (US$)	Names of other agencies involved in project/activity
Determination of typhoon intensity and motion	Regional	To develop methods to determine typhoon motion and intensity		1 person-year		

Major Current/Ongoing Projects/Activities

Title	Geographical coverage (local, national, regional)	Objectives	Generated products	Staff time	Project cost (US$)	Names of other agencies involved in project/activity
Improvement of meteorological information system for better prediction of disasters	National	To improve prediction of meteorological disasters	TPWA and rain rate	3 person-years	360,000	Seoul National University, SERI
Development of an integrated meteorological data system	Regional	To develop a system to integrate various types of meteorological data	Local analysis and display systems	3 person-years	300,000	SERI
Monitoring and predicting typhoon movement through satellites	Regional	To improve forecasting of typhoons using satellites			18,000	KORDI
Using new products from GMS-5	Regional	To develop PWA, UTH, OLR and WVHV	PWA, UTH, OLR and WVHV	2 person-years		

Databases/Archives Developed and/or Maintained and Available to Outside Users

Type (digital or analogue) and storage medium	Database contents (type, location, source data, scale and other useful information)	Geographical extent/surface area in km²	Year of production/ review	Internet address (URL)	Main users
Digital, 8 mm tape	NASA/MSTC SSM/1	Global	1992-1996	yslee@iris.metri.ne.kr	
Digital, 8 mm tape	SMMR T_B		1979-1987	yslee@iris.metri.ne.kr	
Digital, 8 mm tape	SMMR level 1A data		1979-1988	yslee@iris.metri.ne.kr	

R

5.2.14 Singapore

Meteorological Service Singapore

The Meteorological Service Singapore collaborates in WMO with regional and international meteorological agencies in the provision of weather observations, including meteorological satellite data, and in the processing and exchange of meteorological information. Its organization chart is shown in figure 5.3.

Professional Personnel

Primary field of work	Number of staff with the following qualifications			Number of staff with intensive training (>2 consecutive months) during last 5 years
	Ph.D.	MSc	BSc	
Weather forecasting	1	10	7	3 (MSc)
Disaster monitoring/warning				
Environmental monitoring	1	2	2	2 (MSc)
Climate change studies				
Hardware/software development				
Development of ground receiving and analysis stations				
Sensor development				
Radio communications				
Data management/information systems services		1	1	
Management		1		

Total number of professional staff: 26   Total number of administrative staff:

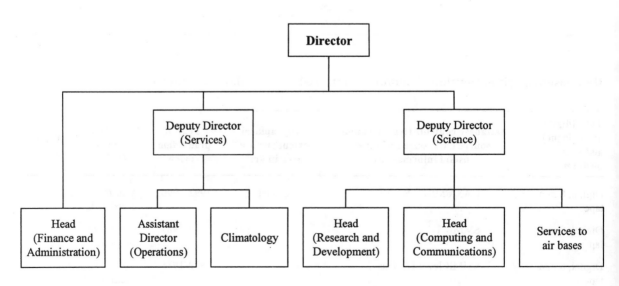

Figure 5.3 Organization chart of the Meteorological Service Singapore

Major Projects/Activities Completed (1992-1996)

Title	Geographical coverage (local, national, regional)	Objectives	Generated products	Staff time	Project cost (US$)	Names of other agencies involved in project/activity
Haze early warning system	ASEAN	To detect and monitor trans-boundary haze	Satellite images, information on sources of haze	1 person-year		ASEAN Specialized Meteorological Centre
GIS on weather	National	To develop a GIS for retrieval of weather information such as rainfall, surface and upper air data over Singapore	Prototype Arc Info-based GIS for weather information over Singapore	0.25 person-year		Nanyang Technological University
Meteorological database processing software for satellite communication	National	To develop software for processing raw meteorological data and presenting them in a form suitable for comparison with data from Earth-space, terrestrial LOC and other radio systems	Software package	0.5 person-year		Nanyang Technological University
Radar and propagation studies in Singapore	National	To analyse received signals from Inmarsat satellites (L/C bands) and Intelsat POR satellites (C band) and investigate signals from Ku-band beacon receiver		0.5 person-years		Nanyang Technological University and Singapore Telecom
Assimilation of ERS-1 surface wind data into limited-area NWP model	Regional	To improve model analysis	Standard NWP products	0.2 person-years		National University of Singapore

Major Current/Ongoing Projects/Activities

Title	Geographical coverage (local, national, regional)	Objectives	Generated products	Staff time	Project cost (US$)	Names of other agencies involved in project/activity
Cloud classification	Regional	To contribute to weather forecasting		0.2 person-year		
Retrieval of sea surface temperature	Regional	To assimilate in NWP models		0.1 person-year		
Retrieval of precipitable water amount	Regional	To assimilate in NWP models		0.1 person-year		
Retrieval of upper tropospheric humidity	Regional	To assimilate in NWP models		0.1 person-year		

S

Databases/Archives Developed and/or Maintained and Available to Outside Users

Type (digital or analogue) and storage medium	Database contents (type, location, source data, scale and other useful information)	Geographical extent/surface area in km²	Year of production/ review	Internet address (URL)	Main users
Digital, optical disks	Binary files	Global, ASEAN, national	1992-1996		
Digital, CD-ROMs	Binary files	Global, ASEAN, national	Since March 1996		

Training Activities (1992-1996)

Course title and description	Training undergone by organization staff		
	Given by (name of institution/agency)	Duration/ frequency	Number of staff taking course
Workshop on haze in ASEAN countries	Ministry of the Environment	2 days	1
Workshop on modelling severe weather	Bureau of Meteorology, Australia	1 week	2
Seminar on tropical cyclone forecasting and research	WMO	2 weeks	1
Course on modelling and forecasting the spread of chemical gases	JICA	4 weeks	1
Seminar on utilization of extended information from the geostationary meteorological satellites	Japan Meteorological Agency	1 week	1

Course title and description	Training provided by organization for external users		
	Name of participating agencies/organizations	Duration/ frequency	Number of participants
Training course on forest fire detection using NOAA meteorological satellite data	Malaysian Meteorological Service	2.5 days	3

5.2.15 Sri Lanka

Department of Meteorology

The Department of Meteorology is a government organization involved in the provision of weather forecasts, especially for aviation and marine purposes; collection and distribution of meteorological data; and provision of climatological and agro-climatological data and consultancy services. Its organization chart is shown in figure 5.4.

Professional Personnel

Primary field of work	Number of staff with the following qualifications			Number of staff with intensive training (>2 consecutive months) during last 5 years
	Ph.D.	MSc	BSc	
Weather forecasting	2	17	2	2 (tropical cyclones and weather forecasting)
Disaster monitoring/warning		6		
Environmental monitoring		2		
Climate change studies	1	3		
Hardware/software development		3		1 (software development for meteorology)
Development of ground receiving and analysis stations				
Sensor development				
Radio communications			1	
Data management/information systems services				
Management				

Total number of professional staff: 25   Total number of administrative staff: 35

Major Current/Ongoing Projects/Activities

Title	Geographical coverage (local, national, regional)	Objectives	Generated products	Staff time	Project cost (US$)	Names of other agencies involved in project/activity
Installation of HRPT receiving network	National	To improve weather forecasting	APT, HRPT images from NOAA, GMS and GOMS		200,000	
Installation of receiving equipment for INSAT cloud imagery	National	To receive enhanced cloud imagery			150,000	Government of India

S

253

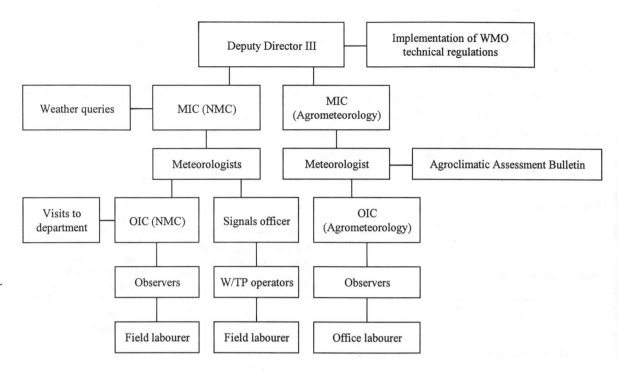

Figure 5.4 Organization chart of the National Meteorological Centre of the Department of Meteorology

Databases/Archives Developed and/or Maintained and Available to Outside Users

Type (digital or analogue) and storage medium	Database contents (type, location, source data, scale and other useful information)	Geographical extent/surface area in km²	Year of production/ review	Internet address (URL)	Main users
Analogue, hardcopy of satellite cloud imagery	NOAA 12, NOAA 14, GMS imagery	40° x 40°	1978 onwards		Department researchers
Analogue, hardcopy of radar imagery	Radar imagery from the weather radar at Trincomalee	400-km radius aroun Trincomalee	1982		

Training Activities (1992-1996)

Course title and description	Training undergone by organization staff		
	Given by (name of institution/agency)	Duration/ frequency	Number of staff taking course
Training course on WF 35 radar	United Kingdom Meteorological Office	2 weeks	1
Meteorological telecom-munications	India Meteorological Office, New Delhi	4 months	2
Satellite meteorology	Meteorological Training Centre, China	4 weeks	1
Microwave remote sensing applications	Satellite Meteorological Centre, China	1 week	1
Software development for meteorology	Meteorological Service Singapore	8 weeks	1

5.2.16 Viet Nam

National Centre for Hydrometeorological Forecasting

The National Centre for Hydrometeorological Forecasting is a government organization engaged in meteorological and disaster monitoring.

Professional Personnel

Primary field of work	Number of staff with the following qualifications			Number of staff with intensive training (>2 consecutive months) during last 5 years
	Ph.D.	MSc	BSc	
Weather forecasting	2	1	4	
Disaster monitoring/warning				
Environmental monitoring				
Climate change studies			1	
Hardware/software development				
Development of ground receiving and analysis stations				
Sensor development				
Radio communications				
Data management/information systems services			2	

Total number of professional staff: 10   Total number of administrative staff: 2

Major Projects/Activities Completed (1992-1996)

Title	Geographical coverage (local, national, regional)	Objectives	Generated products	Staff time	Project cost (US$)	Names of other agencies involved in project/activity
Establishment of meteorological satellite receiving station	Local	To receive high-resolution images from GMS and NOAA satellites	Complete receiving system		1,000,000	Hydrometeorological Service of Viet Nam

Training Activities (1992-1996)

Course title and description	Training undergone by organization staff		
	Given by (name of institution/agency)	Duration/ frequency	Number of staff taking course
Course in satellite meteorology	Hydrometeorological Service of Viet Nam	15 days	45

Course title and description	Training provided by organization for external users		
	Name of participating agencies/organizations	Duration/ frequency	Number of participants
Use of meteorological satellite information in hydrometeorological forecasting	WMO	15 days	3

V

Chapter 6
SATELLITE COMMUNICATIONS AND DISTANCE EDUCATION

6.1 Regional overview

6.1.1 Background

The impact of space technology on the telecommunication sector has been more profound than its impact on other sectors. Space technology has truly revolutionized telecommunications, turning the world into a "global village" through the globe-girdling communication satellite systems that reach out to practically every part of the Earth's surface. Communication satellites offer a wide range of services, including telephony, data transmission, broadcasting, distance education, disaster warning, rural telecommunications, navigation and teleconferencing, among others. Many of the services provided by communication satellites directly affect the everyday lives of ordinary people, in the shape of satellite television and radio, long-distance telephones and data communications, for instance. For most people, therefore, satellite communication is synonymous with space technology.

Satellite communications became commercially viable as far back as the early 1960s; hence the involvement of the commercial sector in this field over the last three decades or more. The operation of the main international satellite communications systems, which handle the bulk of global communication services today, are largely in the hands of intergovernmental organizations, like Intelsat and Inmarsat, which are run on commercial lines. Intelsat is a Washington-based intergovernmental organization founded by 11 member countries in 1964 to develop and provide global satellite communications. By 1995, it had 136 member countries that shared profits in proportion to their use of the system. Intelsat has launched successive generations of satellites with progressively better capabilities. It now has 24 operational satellites in geo-synchronous orbit at 36,000-km altitude over the Atlantic, Pacific and Indian Oceans. Inmarsat, a London-based intergovernmental organization modelled on Intelsat, was founded in 1979. It had 79 member countries by 1995. Inmarsat provides global mobile satellite communications at sea, on land and in the air through its 10 operational geostationary satellites positioned over the three oceans.

Besides Intelsat and Inmarsat, the Asian and Pacific region is served by other global satellite communication systems such as the PanAmSat satellite system which has been operational since 1995. PanAmSat's PAS-2 and PAS-4 satellites provide coverage for the Pacific and Indian Ocean regions respectively. In particular, PAS-4 is being used to beam programmes from a large number of television networks/channels to South Asia and the Middle East. Another international satellite communications system serving the Asia-Pacific region is the Moscow-based Intersputnik, which is also an intergovernmental organization with 22 member countries. It makes use of the Russian Gorizont, Express and Raduga communication satellites.

6.1.2 Regional and national satellite communications system and their uses

In addition to the international systems mentioned above, there are various regional and national satellite communication systems operating in the region. Two of the more well-known are the Hong Kong-based Asiasat and Apstar systems. A large number of television networks belonging to various countries and organizations have leased transponders on these satellites for beaming television programmes. Besides television and radio broadcasting, these satellites are used for supporting VSAT operations and public telephone networks. Examples of national satellite communications systems in the region are Australia's Optus, China's Chinasat, India's INSAT,

Indonesia's Palapa, Japan's JCSAT, Superbird and CS/N-Star, Malaysia's MEASAT, Republic of Korea's Koreasat, and Thailand's Thaicom. INSAT is actually a multi-purpose satellite system providing communication services and meteorological data and supporting a data collection system. Some national systems also service neighbouring countries. For instance, Indonesia's Palapa, which became the world's third national satellite communication system when its first satellite was launched in 1976, has been and/or is used by Malaysia, the Philippines, Singapore and other ASEAN countries, while Australia's Optus is used by New Zealand and Papua New Guinea. These regional and national satellite systems are used for various purposes, including telephony, data communication, direct broadcast television, radio, VSAT operations, mobile communications, distance education and rural telecommunications. They thus complement the services available through Intelsat and other international systems.

Although some national satellite communication systems like China's Chinasat are owned and operated by government agencies, many national systems, such as, for instance, Australia's Optus, Malaysia's MEASAT and Thailand's Thaicom, are owned and managed by companies or groups that are private or joint public-private sector ventures. Many of these operational satellites have been built by well-known aerospace/satellite manufacturers in the United States or Europe. The satellites have been launched mostly by launch service providers in the United States, Europe and China. In line with the growing global and regional trends towards liberalization, privatization and deregulation, more opportunities are being offered by governments to private competitors/service providers to enter the telecommunications sector in their countries. Thus, in many countries the involvement of the private sector in telecommunications, including satellite communications, is expanding with time. Some governments, especially in the developed countries, like Australia, for instance, now want to play a largely supervisory and regulatory role and to be mainly involved in planning the overall development of telecommunications in their countries, leaving much of the operational functions to commercial operators. In particular, the Government of New Zealand has kept a very minimal role for itself with regard to the development of telecommunications in the country. Operational satellite communication systems, which are usually operated by commercial entities, under the regulating authority of ministries/departments concerned with telecommunications, do not normally fall within the purview of activities handled by national space agencies. There are some exceptions, however, such as India, where the national communication-cum-meteorological satellite system, INSAT, is developed, operated and managed by ISRO, the national space agency, rather than by the ministry or department concerned with communications.

Distance education is one of those spin-off benefits of satellite communications that are of special relevance to the developing world. Typically, developing countries have large numbers of illiterate people, especially in the rural areas. The most efficient and economical way of reaching out to the illiterate adults and children across the country is by broadcasting educational television programmes through satellites to a network of community schools or learning centres equipped with suitable VSATs. While such arrangements can be made for both urban and rural areas, they are particularly useful and relevant for the disadvantaged people living in widely scattered villages and settlements across vast rural areas. Apart from basic academic courses, programmes focusing on health, family planning, agriculture, livestock development, vocational skills and other subjects of interest for different target populations can be, and are being, beamed through satellites. Some countries where satellites are being extensively used for distance education are Australia (mainly for indigenous Aborigines), China, India, Indonesia, the Philippines and Thailand. Many other countries are also beginning to use communication satellites for broadcasting educational programmes. Various projects are also being implemented in different regional countries specifically to improve rural telecommunications as such, using communication satellites, as satellites offer a quick and cost-effective way of achieving these objectives. Communication satellites are also used to relay information on impending disasters like typhoons, cyclones and floods and issue advance warnings and alerts so that necessary precautionary measures may be taken in time.

In view of its wide-ranging utility, satellite communications has been selected as one of the four application areas on which attention is being focused during the implementation of the ESCAP Regional Space Applications Programme for Sustainable Development (RESAP). Accordingly, a Regional Working Group on Satellite Communication Applications has been formed under RESAP. At its second meeting, held in Bali, Indonesia, in March 1997, the Working Group identified three priority areas for its medium-term plan, namely distance education, information superhighway infrastructure and rural communications. All these priority areas are pertinent to the needs of the Asia-Pacific region, and any collaborative projects carried out in these areas should prove very beneficial to the region.

6.1.3 Future developments in satellite communications

The range of satellite-based services in the region is expected to expand as more national and regional satellite communication systems become operational over the next few years. These include, besides further satellites in existing national systems like INSAT and Thaicom, several new systems, which include Indonesia's Indostar, the Philippines' Agila and Mabuhay satellites, a Singapore satellite system and some regional systems offering mobile satellite communication services, namely, Garuda, Agrani and APMT. The last three systems are financed by various private companies or consortia mainly based in the region. Further, the Islamic Republic of Iran's Zohreh and Pakistan's Paksat satellite communications systems are also planned to be developed in the coming years.

There is a growing global trend towards providing more mobile satellite communications services. A number of international satellite communications systems that will mainly deploy low Earth orbit (LEO) and/or medium Earth orbit (MEO) satellites are being planned, designed and developed, such as Iridium, GlobalStar, Odyssey and Inmarsat I-CO. They will offer global direct mobile satellite communication services, thereby greatly extending the scope of satellite-based personal communications. Although these large global systems are being designed and fabricated in developed countries outside the region, the planners and investors have taken into consideration the rapidly increasing demand for satellite communications, including mobile services, in the Asia-Pacific region with its huge population and rising prosperity. These global systems are being handled or financed by different consortia or groups, led by well-known international and American companies and with investments by various groups and investors across the world, including several from the Asia-Pacific region. The Asia-Pacific region, with its rising prosperity and increasing demand, will undoubtedly be a large and growing market for these global satellite mobile communication systems.

6.1.4 Conclusion

The Asia-Pacific region is well served by international and regional as well as national satellite communication systems as more and more countries in the region invest in their own national satellites. The infrastructure relating to satellite communications in the various countries continues to be upgraded and expanded, and there is a definite shift towards greater involvement of private service providers in the development of telecommunications. Satellite communication systems are being used for various services and functions, including distance education, especially for people living in the countryside. Satellite communications can thus play a significant role in the social and economic uplift of less fortunate groups of people as well as in overall national economic development. With new and more versatile satellite systems being designed and launched, the impact of satellite communications on national socio-economic development in the countries of the region will be even greater in the coming years. Undoubtedly, there is a bright future in the region for satellite communications.

Table 6.1 National contact points of the Regional Working Group on Satellite Communication Applications and Distance Education

Country	Name	Agency
Australia	Professor Michael Miller	University of South Australia
Azerbaijan	–	–
Bangladesh	–	–
China	Mr Wang Yan Guang	Beijing Institute of Statellite Information Engineering
Fiji	Mr Bruce Davies	University of South Pacific
India	–	–
Indonesia	Mr Lukman Hutagalung	Directorate General of Post and Communications
Islamic Republic of Iran	Mr Mahammad Hakak	Iran Telecommunication Research Centre
Japan	Mr Masakuni Esaki	Ministry of Posts and Telecommunications
Malaysia	Mr Mohd. Aris Bernawi	Ministry of Energy, Telecommunications and Post
Mongolia	Mr Sanjaa Ganbaatar	Mongolian Telecommunications Company
Myanmar	–	–
Nepal	Mr Bhesh Raj Kanel	Telecommunications Training Centre
Pakistan	Mr S.A.H. Abidi	Space and Upper Atmosphere Research Commission
Philippines	Ms Aurora A. Rubio	Department of Transportation and Communications
Republic of Korea	–	–
Russian Federation	–	–
Singapore	Mr Tan Soon Hie	Nanyang Technological University
Sri Lanka	Director	Arthur C. Clarke Centre for Modern Technologies
Thailand	Mr Kosol Choochuay Mr Rasamee Suwanwerakamthorn	Ministry of Education National Research Council (alternate)
Vanuatu	–	–
Viet Nam	Mr Nguyen Than Phuc	Science-Technology and International Cooperation

Table 6.2 Existing facilities/infrastructure relating to satellite communications

Agency/country	Computer			Satellite systems currently being used	Satellite systems planned	Network	Software
	MF	WS	PC				
Azerbaijan National Space Agengy, Azerbaijan			✓	Intelsat, Eurosat, Turksat, Horizon			Self-developed
National Remote Sensing Centre, Changsha, China			✓				
Wuhan Technical University of Surveying and Mapping, China		✓	✓			Internet, Fast Ethernet	

Table 6.2 *(Continued)*

Agency/country	Computer			Satellite systems currently being used	Satellite systems planned	Net-work	Soft-ware
	MF	WS	PC				
Telecommunication Advancement Organization of Japan	✓	✓		JCSAT-3		Internet, Ethernet	Windows NT, Oracle, Borland C++
Mongolian Telecommunications Company, Mongolia			✓	Inter-sputnik, Intelsat, Asiasat	Inmarsat	Novell, Internet	Windows95, Novell Netware
Nepal Telecommunications Corporation, Nepal			✓	Intelsat		LAN, Internet	VSAT, DACOS, DCME, MAC
Space and Upper Atmosphere Research Commission, Pakistan	✓		✓	Amsat, Kitsat, UoSat, PCSAT		LAN	
Nanyang Technological University, Singapore	✓		✓	Intelsat, Inmarsat, Palapa, Asiasat, UoSat, Kitsat, PCSAT	UoSat-12	Internet, Ethernet	PCC, TLM, PDB
Arthur C. Clarke Centre for Modern Technologies, Sri Lanka	✓	✓	✓	Intelsat, amateur satellites	Same	LAN, Internet	
Thaicom Distance Education Centre, Thailand			✓	Ku-band direct-to-home broadcasting	Same	LAN, Internet	
Post and Telegraph Department, Thailand	✓		✓	Thaicom, Intelsat, Inmarsat	Thaicom-4, Iridium	Internet	Stream-line for TDMA/DAMA
Telephone Organization of Thailand				Thaicom	Thaicom 2/3		
Communication Authority of Thailand		✓	✓	Thaicom, Intelsat, Inmarsat	Same	LAN, PPP, Internet	Satellite communication control

Table 6.3 Estimated investment (US dollars) in infrastructure

Country	Agency	1992-1996	1996-2000
China	Wuhan Technical University of Surveying and Mapping	130,000	68,000
China	National Remote Sensing Centre, Changsha	95,000	120,000
Japan	Telecommunication Advancement Organization of Japan	9,370,000	–
Mongolia	Mongolian Telecommunications Company	11,000,000	8,000,000
Nepal	Nepal Telecommunications Corporation	2,500,000	2,500,000
Pakistan	Space and Upper Atmosphere Research Commission	1,500,000	3,000,000
Singapore	Nanyang Technological University	300,000	3,000,000
Sri Lanka	Arthur C. Clarke Centre for Modern Technologies	15,000	–
Thailand	Thaicom Distance Education Centre	85,000,000	65,000,000
Thailand	Post and Telegraph Department	4,400,000	13,000,000
Thailand	Communications Authority of Thailand	6,900,000	93,600,000

6.2 Country inputs

6.2.1 Azerbaijan

Azerbaijan National Space Agency

The Azerbaijan National Space Agency (ANASA) is a government organization whose mandate is specified in the decree issued by the President of Azerbaijan.

Professional Personnel

Primary field of work	Number of staff with the following qualifications			Number of staff with intensive training (>2 consecutive months) during last 5 years
	Ph.D.	MSc	BSc	
Rural telecommunications: education				
Rural telecommunications: health care				
Broadcasting of meteorological information: weather forecasts	1	10	40	
Broadcasting of meteorological information: disaster warnings and relief	1	3	20	
Hardware/software development	2	10	30	
Development of ground terminals (TVROs, VSATs, portable terminals)		5	20	
Data management/information systems services	1	7	25	

Total number of professional staff: 174

Total number of administrative staff: N/A

Major Current/Ongoing Projects/Activities

Type of electronic media used	Subject/topic	Co-sponsorship (if any)	Nature and size of intended public
	Creation of satellite communication system in Azerbaijan	Ministry of Communications	The entire population of Azerbaijan

Training Activities (1992-1996)

Course title and description	Training provided by organization for external users		
	Name of participating agencies/organizations	Duration/ frequency	Number of participants
Earth study from space; aerospace device development; measure-information systems; automatic systems for information processing and control	Azerbaijan State Oil Academy, State Technology University	Annual	70

Electronic Media Programmes Developed since 1990 and Available to Outside Users

Type (radio, television, computer network, other)	Programme contents (subject and other relevant information)	Year of production	Intended public
Computer network	Programme set for thematic mapping of underlying surfaces taking into account optical inhomogeneities in atmosphere	1995	Ministry of Communications Committee on Hydrometeorology

6.2.2 India

FACILITIES AND ACTIVITIES*

Introduction

The beneficial role of satellite-based systems for education and development communications was recognized in India in the 1960s. A number of studies were conducted to identify a suitable system to meet the requirements of India with its varied regional, cultural and linguistic requirements. A path-finding experiment in the use of satellite broadcasting to provide rural communities with information that would be useful in their day-to-day life was conducted in 1975-1976 using the ATS-6 satellite. Under this experiment called SITE (Satellite Instructional Television Experiment), programmes in local languages on agriculture, health, animal husbandry and family welfare were broadcast directly for community viewing through dish antennae in 2,400 villages. SITE also had a school broadcast component with enrichment programmes for children in the primary school age group, which included science programmes specially designed for rural children and relating to their immediate environment.

The success of SITE led to the establishment of the operational multi-purpose Indian National Satellite system, that is, INSAT, in the early 1980s. INSAT provides telecommunications, broadcasting and meteorological services. Presently, the INSAT system consists of four operational satellites, namely INSAT-lD (83ºE), INSAT-2A (74ºE), INSAT-2B and INSAT-2C (collocated at 93.5ºE). Together they provide over 60 transponders for telecommunication and television services. These services are expected to be further augmented by INSAT-2D launched in June 1997 and by INSAT-2E scheduled for launch in 1998.

The satellite television component consists of 18 television channels, which include national and regional language services. These are re-broadcast through the terrestrial television transmitters. Cable operators also distribute them through cable. The terrestrial television network consists of over 750 transmitters covering 85 per cent of the population and 68 per cent of the area of the country.

Presently, in India, satellite-based education, development communications and training activities have three components:

(a) Limited-duration broadcasts over the national and regional television networks for schoolchildren, university students, adult education and rural development;

(b) An interactive distance education, training and development communication channel providing tele-education and training for different rural development functionaries, students and counsellors of open universities, primary teachers, professional students, industrial workers and bank staff, among others;

(c) An area-specific dedicated rural development communication channel for the Jhabua Development Communications Project in a backward district of Madhya Pradesh state, which is a pilot project designed to provide inputs for implementation of a large programme on a wider scale.

A. Educational television broadcasts

1. Primary school

INSAT is used to provide an educational television (ETV) service for primary schoolchildren in selected three-district clusters in the six states of Andhra Pradesh (Telugu), Orissa (Oriya), Maharashtra (Marathi), Gujarat (Gujarati), Bihar (Hindi) and Uttar Pradesh (Hindi). At present,

---

*Based on a report provided by the Indian Space Research Organization.

educational programmes put out from Delhi via the S-band broadcasting transponder of INSAT-lD are relayed through all television relay transmitters in Bihar, Uttar Pradesh, Gujarat and Orissa daily in a "time-sharing" mode. The ETV programmes in Hindi meant for Uttar Pradesh and Bihar are also relayed by all transmitters in Madhya Pradesh, Rajasthan, Haryana and Himachal Pradesh. The Marathi and Telugu ETV programmes are telecast from Maharashtra and Andhra Pradesh, respectively.

ETV programmes are meant to provide general enrichment of the mind for schoolchildren and are not related to any specific school curriculum. The duration of these programmes in each language is 45 minutes. Separate programmes for primary schoolchildren in the age groups of 4 to 8 years and 9 to 11 years are also telecast. The ETV programmes are targeted for rural schools where television brings teaching aids which are otherwise unavailable or scarce to the classrooms. Programmes for the guidance/training of primary schoolteachers are also telecast. Responsibility for the production of these programmes rests with the educational authorities at the central and state levels.

Various educational institutions are engaged in the production of television programmes for telecasting on the ETV channel. The content of the programmes is chosen with the following objectives in mind:

(a)   To support and extend the existing curriculum in environmental science, social sciences, languages and mathematics;

(b)   Enrichment by including direct values, good habits, national awareness;

(c)   Programmes for teachers to enhance their competence in the subject areas, methods of teaching and teaching aids.

2. Higher education

A one-hour general enrichment programme on higher education (college sector) is telecast on the national network. These programmes provided by the University Grants Commission are in English and are a part of its country-wide classroom programme.

The programmes are selected with the following objectives:

(a)   To put quality education within the reach of students in small villages and towns;

(b)   To level the disparities caused by lack of facilities, good teachers and teaching aids in many colleges in the country;

(c)   To remove obsolescence in textbooks and to acquaint teachers and students with the latest developments in different disciplines.

The Indira Gandhi National Open University broadcasts curriculum-based lectures for the students of the open university for half an hour daily via the national network. These broadcasts are curriculum-based and delivered in English.

B. Formal distance education

1. Tertiary level

In India, over the past two decades, the formal distance education system at the tertiary level has grown significantly, with about 50 distance education institutions in the country. Apart from the correspondence course institutes, the distance education system in India also encompasses the open university system. There are presently 51 universities, including deemed universities, offering correspondence education programmes in the country.

The Indira Gandhi National Open University was founded on 20 September 1985 by an Act of Parliament. It is responsible for the promotion, maintenance of standards and coordination of open and distance education system in India in addition to functioning as a university for open learning and distance education programmes. To carry out its responsibility as an apex body for distance education in India, it established the Distance Education Council. This oversees the promotion of distance education system in the country, coordination and maintenance of the standards in distance education, and provides financial assistance to state open universities.

There are six state-level open universities in India. These are Dr B.R. Ambedkar Open University, Hyderabad, AP (started in 1982); Nalanda Open University, Patna, Bihar (1987); Kota Open University, Kota, Rajasthan (1987); Yashwantrao Chavan Maharashtra Open University, Nasik, Maharashtra (1989); M.P. Bhoj University, Bhopal, Madhya Pradesh (1992); and Dr Babasaheb Ambedkar Open University, Ahmedabad, Gujarat (1994). The present enrolment in the distance education system in India is around one million, which constitutes about 14 per cent of the total enrolment at the tertiary level in the country. The number is expected to triple by the turn of the century.

2. Secondary level

Correspondence courses at secondary level were started in 1965 by the Board of Secondary Education, Madhya Pradesh (now referred to as M.P. Open School), with the objectives of improving standards of those who have dropped out from the school system and enabling them to complete their secondary education. This was followed by five states (Delhi, Rajasthan, Orissa, Tamil Nadu and Uttar Pradesh) which established institutions offering correspondence education at secondary and higher secondary levels. The first open school offering secondary courses was established in New Delhi in 1979. This open school was upgraded to national open school in November 1989, with the objective of providing relevant, continuing and developmental education to prioritized client groups as an alternative to the formal system. The states of Punjab and Andhra Pradesh have also established open schools recently. The national and state open schools together have at present an enrolment of about 260,000 students.

3. Technologies

In distance education, the students and the teacher are at a distance from one another with little opportunity for face-to-face contact. The bulk of the learning is done through self-study, at times and places convenient for the students. Also, the nature of the distance education system is such that it requires the instructional material, delivery system and support services to be designed in a way which facilitates independent learning. The multi-media approach encompassing print, audio/video cassettes, television/radio broadcasts, computer-assisted instruction, counselling, etc., is already being followed in various distance education institutes.

At present, distance educational institutions rely heavily on the local expertise of academic counsellors to help learners in their self-study and on print materials. To meet the ever-increasing load on the system, distance education is looking for appropriate and innovative technologies for mass communications. The satellite media are being increasingly used to meet the requirements of distance education. Apart from broadcasts via INSAT, an interactive satellite system is being used by the Open University to reach its students and to train the distance education counsellors.

C. Interactive satellite communication systems

The addition of an interactive component to the television broadcast can considerably enhance its effectiveness for developmental communications and training. In the first half of the 1990s, several demonstrations were carried out by Indian Space Research Organization (ISRO) via INSAT in the use of a satellite-based interactive communication system (one-way-video and two-way-audio) in several application areas.

In this interactive system, the "teaching-end" comprises a studio where the experts deliver, either live or through pre-recorded tapes, lectures on the subject. These lectures in television form (video and audio) are transmitted to the satellite through a large Earth station which is linked or collocated with the studio. The satellite relays back the television signals for reception by participants at classrooms dispersed across the country directly by small satellite dish antennas. Participants/students in the "classrooms" can ask questions to the experts present at the "teaching-end" on an audio channel through dish antennas modified to include audio transmission capability. At the "teaching-end" the questions received from a "classroom" are looped back on the audio channel of the television signal emanating from the "teaching-end" so that the questions can be heard at all "classrooms". The response to the questions by the expert goes on the television signal and is received by all classrooms. This mode of interaction essentially simulates the learning environment of a conventional classroom, thereby creating a virtual classroom. Some of the "classrooms" which are not equipped with satellite talk-back terminals can ask the questions to the teaching end on normal long-distance telephone lines.

This system is particularly useful for simultaneously training large numbers of people at large numbers of dispersed locations. It has been used to train adult education functionaries, rural development functionaries, "panchayatiraj" women elected members, students of engineering and management, industrial workers and banking staff, among others.

Some of the advantages of the system are:

- Training a large number of people who are geographically dispersed in the shortest time

- Multiplier effect due to training the trainers; uniformity of training content

- Access to the best available learning resources, irrespective of the geographic location of the learners

- Repeatability of training courses/educational packages; easy updating and dissemination

- Enhanced involvement of the trainees/learners due to interaction capability and therefore greater learning gains; enterprise-wide participation

- The ability to share the same network by different user groups; specific topics for specific locations also possible

- Significant savings in expenditure due to economies in travel, logistics and replication of teaching infrastructure; more frequent training

- It can effectively supplement the conventional system of training and in some cases can itself become the major component of the training system

D. Training and development communication channel

The response to the demonstrations of the interactive system were so overwhelming that an operational system was established in 1995. An extended C-band transponder has been earmarked as the training and development communication channel (TDCC) on INSAT for this purpose. A Government of India Secretary-level inter-departmental Steering Committee provides policy guidelines for the utilization and growth of this channel. To allow the users to avail themselves of the channel, uplink and studio facilities have been set up in two locations, one in New Delhi at the Indira Gandhi National Open University campus and the other in Ahmedabad at the ISRO campus. These facilities, which constitute high-cost components in the network, are being shared by several user agencies by paying hourly utilization charges. The users will be responsible for establishing the receiving terminals, organizing the classrooms, course content, resource persons and video production. Utilization of the channel has grown substantially over the past two years. From an average of ten days in a month in 1995, the present utilization is about 22 days a month.

This channel is being extensively used by the state governments for training primary teachers, "panchayatiraj" (local bodies) elected representatives, "anganwadi" workers associated with women and child development, watershed development functionaries, health and family welfare functionaries, animal husbandry and cooperative members. The number of personnel receiving training is very large, making this system the most appropriate one. For example, the number of "panchayatiraj" elected representatives, "anganwadi" workers and primary teachers runs into hundreds of thousands.

The All India Management Association, the Institute of Electronics and Telecommunications Engineers use this channel regularly for conducting courses and training counsellors. The National Council of Educational Research and Training uses this channel for training primary teachers. The utilization of the system by individual organizations has gone up substantially. Many organizations plan to establish their own dedicated uplinks and studios in their premises.

E. Rural development: Jhabua development communications project

It is well accepted that the value of developmental communications programmes is enhanced if broadcasts are in the local language and address locale-specific issues. It is in this context that the concept of Gramsat Network for broadcasting directly to the rural areas, area-specific developmental/educational/training programmes in the local language, has been conceived. "Gram" in Hindi means village. The conception and establishment of an operational development communication network encompassing over 450 districts in the country calls for innovations in the technological, managerial, programming and social dimensions. It includes the establishment of the satellite television transmitting and receiving network consisting of large numbers of TVROs in rural areas and maintenance of the same. The television sets should be accessible to the target audience, i.e., the community mode of reception is to be provided. Identifying the area-specific communication needs of the rural areas and linking them with ongoing development activities under implementation is a major effort. The production of programmes calls for a massive effort and trained manpower as these programmes have to be made not only informative but also attractive for the target audience. Simultaneously, the costs have to be kept low. The involvement of people and district officials and voluntary organizations in all stages of the system, from defining requirements to follow-up and feedback, is crucial for its success.

To demonstrate the efficacy of a satellite-based development communication and training network for rural development, a pilot project, namely the Jhabua Development Communications Project (JDCP), is being carried out in India in the Jhabua District of Madhya Pradesh for a period of two years. Regular transmissions started on 1 November 1996. This "end-to-end" project demonstrates the effectiveness of communication support to the developmental activities in the district and also in providing interactive training to field officials and the people in general. It will also provide inputs for planning and establishing operational systems on a wider scale in the country.

Jhabua is one of the most backward districts in the country and has the highest tribal population, at 85 per cent. The literacy level is low at 14 per cent. Dialects are used in the district, but people understand simple Hindi. In this project 150 receiving terminals at the village level and one talkback terminal in each of the 12 block headquarters are installed. This network of talkback and receiving terminals is utilized to conduct training programmes for the field staff and for communicating specific development-oriented messages to the audiences at the receiving terminals. The studio facilities and ISRO Earth station in Ahmedabad are being used as the teaching end.

The priority areas of development where communication support is required include watershed management, health, education and "panchayatiraj". Watershed development includes agriculture, animal husbandry, forestry and fisheries. The content of the programmes to be transmitted is defined jointly with subject experts and state/district/field officials, keeping the needs of the people of Jhabua in view. While the latest inputs are taken from national-level resource persons and institutions, the training material is produced in the field, with the people, so that it can be easily comprehended. Detailed feedback and evaluation is carried out for continuous improvement and overall assessment of the impact.

F. Future programmes

TDCC and JDCP are just the beginning of a concerted effort to use space technology to meet the communication and training requirements of distance education and rural areas. TDCC is likely to grow in the near future with more transponders and uplinks to cover larger user bases and to spread geographically. Replication of JDCP in other backward districts of the country is also being considered. A satellite-based open education network for the education and training of a large number of students, in-service education of primary teachers to cover three million primary teachers, and the Indian Training and Education Network for Development are also being conceptualized. The future network expansions will be based on judicious use of the advances that are taking place in digital technologies.

G. Contact addresses

Training and development communication channel and Jhabua development communications project

Director
Satellite Communications Programmes Office
ISRO
Antariksh Bhavan, New B.E.L. Road
Bangalore 560 094, India
Phone: +91-80-3415301
Fax: +91-80-3412141

Director
Development and Educational Communications
Unit
ISRO
SAC PO, Ahmedabad 380 053, India
Phone: +91-79-6743954
Fax: +91-79-6568556

Formal distance education

Open Universities and Distance Education Council

Vice Chancellor
Indira Gandhi National Open University
Maidan Garhi
New Delhi 110 068, India
Phone: +91-11-6862707
Fax: +91-11-6862312

University Grants Commission: country-wide classroom

Director
Consortium for Educational Communication
NSC Campus, Aruna Asaf Ali Marg
New Delhi 110 067, India
Phone: +91-11-6896637
Fax: +91-11-6897416

School programmes

Director
National Council of Educational Research
 and Training
Aurobindo Marg
New Delhi 110 016, India
Phone: +91-11-6864801
Fax: +91-11-6864141

I

6.2.3 Indonesia

Ministry of Posts, Tourism and Telecommunication, Directorate of Frequency Management

The Directorate of Frequency Management of the Ministry of Posts, Tourism and Telecommunication conducts activities in the field of radio frequency and satellites based on technical policies formulated by the Director General of Posts and Telecommunications.

Professional Personnel

Primary field of work	Number of staff with the following qualifications			Number of staff with intensive training (>2 consecutive months) during last 5 years
	Ph.D.	MSc	BSc	
Rural telecommunications: education				
Rural telecommunications: health care				
Broadcasting of meteorological information: weather forecasts				
Broadcasting of meteorological information: disaster warnings and relief				
Hardware/software development			1	4
Development of ground terminals (TVROs, VSATs, portable terminals)				
Data management/information systems services		1	3	40 (frequency management systems)

Total number of professional staff: 25   Total number of administrative staff: 75

Major Current/Ongoing Projects/Activities

Type of electronic media used	Subject/topic	Co-sponsorship (if any)	Nature and size of intended public
Radio	Radio monitoring system	France	Internal (within the Ministry)
Computer network	Automated frequency management systems	Canada	Internal (within the Ministry)
Internet, fax and phone	Coordination of the regional working group on satellite communication applications		ESCAP and participating countries in the ESCAP region
Fax and phone	World radio communication conference		ITU
Satellite	Satellite coordination meetings		ITU-Members
	Frequency coordination meetings		General public in the country, national level
	Frequency channelling plan		General public in the country, national level
	Frequency assignment		General public in the country, national level
	Frequency controlling		General public in the country, national level
	Frequency allowance plan		–

Training Activities (1992-1996)

Course title and description	Training undergone by organization staff		
	Given by (name of institution/organization)	Duration/ frequency	Number of staff taking course
Radio monitoring system training	THOMSON-CSF	3 months	21
Automated frequency management systems	SPECTROCAN	2 months	40
Satellite coordination assistance	Radiocommunication Bureau	1 year	5
Spectrum management training	SMA Australia and New Zealand	3 months	1

Course title and description	Training provided by organization for external users		
	Name of participating agencies/organizations	Duration/ frequency	Number of participants
Frequency management	Department General of Posts and Telecommunications, Viet Nam	1 week	6
Radio monitoring system training	Regional Office of Department of Tourism, Posts and Telecommunications	1 month	21

Electronic Media Programmes Developed since 1990 and Available to Outside Users

Type (radio, television, computer network, other)	Programme contents (subject and other relevant information)	Year of production	Intended public
Television	How to make the community aware of using frequency wisely	1992	General public in the country
Computer	Radio frequency management	1996	Internal (within the Ministry)

Written Documents Published since 1990

Title	Year of publication	Language of publication	Keywords	Frequency
Directorate of frequency management leaflet	1996	Indonesian		
Radio frequency monitoring system leaflet	1995	Indonesian		

6.2.4 Islamic Republic of Iran

Iran Telecommunication Research Centre

The Iran Telecommunication Research Centre (ITRC) conducts research activities in telecommunication on the basis of national needs and also acts as the authority in drafting and prescribing telecommunication standards.

Professional Personnel

Primary field of work	Number of staff with the following qualifications			Number of staff with intensive training (>2 consecutive months) during last 5 years
	Ph.D.	MSc	BSc	
Rural telecommunications: education				
Rural telecommunications: health care				
Broadcasting of meteorological information: weather forecasts				
Broadcasting of meteorological information: disaster warnings and relief				
Hardware/software development	1	3	4	
Development of ground terminals (TVROs, VSATs, portable terminals)	2	6	10	
Data management/information systems services		1	2	
Telecommunications	5	25	60	
Electronics	3	10	30	
Computers	1	15	40	

Total number of professional staff: 393   Total number of administrative staff: 120

Major Current/Ongoing Projects/Activities

Type of electronic media used	Subject/topic	Co-sponsorship (if any)	Nature and size of intended public
	Cellular telecommunication system (PCS)		
	Digital microwave radio		
	SCPC Earth station		
	Multiplexing equipment		
	Optical telecommunication system		
	Data communication equipment		
	Information technology		
	Standardization of telecommunication equipment		

Training Activities (1992-1996)

Course title and description	Training undergone by organization staff		
	Given by (name of institution/organization)	Duration/ frequency	Number of staff taking course
Operation and maintenance of automatic test equipment	ITRC, in-house training for organization's staff		20
Operation and maintenance of low capacity switch	ITRC, in-house training for organization's staff		25
Microcontroller	ITRC, in-house training for organization's staff		25
Optical telecommunications	ITRC, in-house training for organization's staff		10
Artificial intelligence	ITRC, in-house training for organization's staff		15

Course title and description	Training provided by organization for external users		
	Name of participating agencies/organizations	Duration/ frequency	Number of participants
Operation and maintenance of automatic test equipment	Telecommunications Company of Iran		20
Operation and maintenance of low capacity switch	Telecommunications Company of Iran		20

I

Satellite Communication Department, Telecommunication Company of Iran

The Satellite Communication Department of the Telecommunication Company of Iran is a government organization which provides local, long-distance and international services for the public and private sectors in the Islamic Republic of Iran. More information about the company may be obtained at their Web site, <http://www.dciweb.dc.co.ir>.

Training Activities (1992-1996)

Course title and description	Training undergone by organization staff		
	Given by (name of institution/organization)	Duration/ frequency	Number of staff taking course
Introduction to VSAT	HNS	4 weeks	15
Advanced theory of data communications by satellite	HNS	4 weeks	15
ITU course on coordination of satellite systems	ITU	1 week	12
Mobile satellite systems	Inmarsat	1 week	14
Earth station technology	Alcatel Space	2 weeks	20

Documents Published since 1990

Title	Year of publication	Language of publication	Keywords	Frequency
Role of GMPCS in the future of telecommunications	1996	Persian		
Satellite and Earth station technology	1996	Persian		
Coordination of satellite systems	1996	Persian		

6.2.5 Japan

Telecommunication Advancement Organization of Japan

The Telecommunication Advancement Organization of Japan is a semi-autonomous government organization which conducts research on human resources development and network technologies.

Professional Personnel

Primary field of work	Number of staff with the following qualifications			Number of staff with intensive training (>2 consecutive months) during last 5 years
	Ph.D.	MSc	BSc	
Rural telecommunications: education	1	2		
Rural telecommunications: health care				
Broadcasting of meteorological information: weather forecasts				
Broadcasting of meteorological information: disaster warnings and relief				
Hardware/software development				
Development of ground terminals (TVROs, VSATs, portable terminals)				
Data management/information systems services				

Total number of professional staff: 12   Total number of administrative staff: 6

Major Current/Ongoing Projects/Activities

Type of electronic media used	Subject/topic	Co-sponsorship (if any)	Nature and size of intended public
Video, CAI	Development of human resources for telecommunications		

Training Activities (1992-1996)

Course title and description	Training provided by organization for external users		
	Name of participating agencies/organizations	Duration/ frequency	Number of participants
Data transmission basis	King Mongkut Institute of Technology (Thailand) and Nanking University (China)		60 (30 from each agency)
Communication network	King Mongkut Institute of Technology (Thailand) and Nanking University (China)		60 (30 from each agency)
OSI	King Mongkut Institute of Technology (Thailand) and Nanking University (China)		60 (30 from each agency)
Terminal, line, centre equipment design	King Mongkut Institute of Technology (Thailand) and Nanking University (China)		60 (30 from each agency)

J

6.2.6 Mongolia

Mongolia Telecommunications Company

The Mongolia Telecommunications Company is a company with shares owned by the Government of Mongolia and Korea Telecom. It is the main provider of national and international telecommunication services in Mongolia and has the monopoly in this field. Its organization chart is shown in figure 6.1.

Professional Personnel

Primary field of work	Number of staff with the following qualifications			Number of staff with intensive training (>2 consecutive months) during last 5 years
	Ph.D.	MSc	BSc	
Rural telecommunications: education				
Rural telecommunications: health care				
Broadcasting of meteorological information: weather forecasts				
Broadcasting of meteorological information: disaster warnings and relief				
Hardware/software development			6	
Development of ground terminals (TVROs, VSATs, portable terminals)			15	
Data management/information systems services			2	
Earth station operations			20	
Maintenance			3	
Network engineering	1			

Total number of professional staff: 37   Total number of administrative staff: 17

Training Activities (1992-1996)

Course title and description	Training undergone by organization staff		
	Given by (name of institution/organization)	Duration/ frequency	Number of staff taking course
Principles of data communications	C and W Telecommunications College	1 month	1
Satellite communications	C and W Telecommunications College	1 month	1
Satellite communications engineering	JICA	1-2 months	6

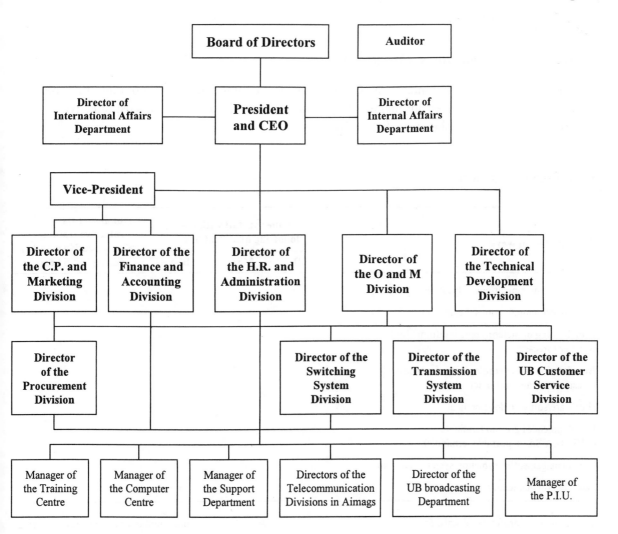

Figure 6.1 Organization chart of the Mongolian Telecommunications Company

M

6.2.7 Nepal

Nepal Telecommunications Corporation

The Nepal Telecommunications Corporation is an autonomous corporation that provides telecommunications services throughout Nepal in a practical, reliable and cost-effective manner. Its organization chart is shown in figure 6.2.

Professional Personnel

Primary field of work	Number of staff with the following qualifications			Number of staff with intensive training (>2 consecutive months) during last 5 years
	Ph.D.	MSc	BSc	
Rural telecommunications: education				
Rural telecommunications: health care				
Broadcasting of meteorological information: weather forecasts				
Broadcasting of meteorological information: disaster warnings and relief			1	
Hardware/software development				
Development of ground terminals (TVROs, VSATs, portable terminals)				
Data management/information systems services				
Satellite communications			4	1

Total number of professional staff: 5   Total number of administrative staff: 20

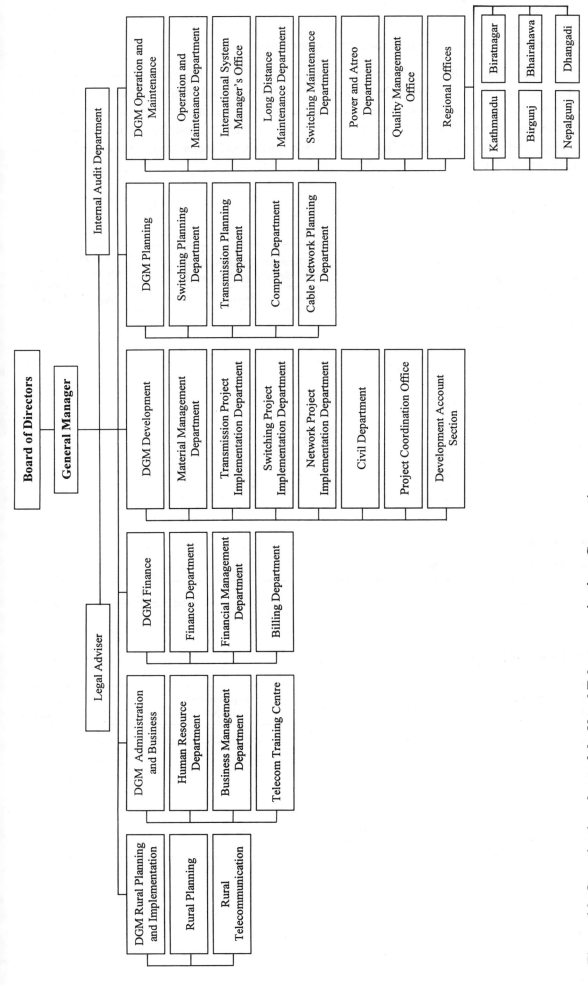

Figure 6.2 Organization chart of the Nepal Telecommunications Corporation

N

6.2.8 Pakistan

Space and Upper Atmosphere Research Commission

The Space and Upper Atmosphere Research Commission (SUPARCO) is an autonomous commission under the federal government. As the national space agency, SUPARCO is responsible for the implementation of the national space programme, which is essentially aimed at furthering research in space science and related fields, enhancing capabilities in space technology and promoting the applications of space science and technology for the socio-economic development of the country.

Professional Personnel

Primary field of work	Number of staff with the following qualifications			Number of staff with intensive training (>2 consecutive months) during last 5 years
	Ph.D.	MSc	BSc	
Rural telecommunications: education				
Rural telecommunications: health care				
Broadcasting of meteorological information: weather forecasts				
Broadcasting of meteorological information: disaster warnings and relief				
Hardware/software development	1	4	6	
Development of ground terminals (TVROs, VSATs, portable terminals)		2	4	
Data management/information systems services		2	3	
System planning and management	1	2	2	
COSPAR/SARSAT programme		1	2	
Frequency and coordination planning		1	1	

Total number of professional staff: 50   Total number of administrative staff: 10

Major Current/Ongoing Projects/Activities

Type of electronic media used	Subject/topic	Co-sponsorship (if any)	Nature and size of intended public
Satellite	Satellite-aided search and rescue in Pakistan	COSPAS-SARSAT parties	National and international users
	Design and development of satellite systems and subsystems		National
	Design and development of small ground terminals		National

Electronic Media Programmes Developed since 1990 and Available to Outside Users

Type (radio, television, computer network, other)	Programme contents (subject and other relevant information)	Year of production	Intended public
Radio	Satellite-aided search and rescue in Pakistan	1991	Users of emergency beacons and information
Radio	Pakistan's satellite programme	1991	National academic institutions

Training Activities (1992-1996)

Course title and description	Training undergone by organization staff		
	Given by (name of institution/organization)	Duration/ frequency	Number of staff taking course
Course on satellite communications arranged under TOKTEN (transfer of knowledge through expatriate nationals) programme	Carleton University, Canada	3 weeks	15
Hands-on training on data communication networks arranged under TOKTEN programme	Raytheon, United States	3 weeks	12

Course title and description	Training provided by organization for external users		
	Name of participating agencies/organizations	Duration/ frequency	Number of participants
Training course on utilization of COSPAS-SARSAT system and data	Various national agencies such as CAA, NSC, PTS and other users of COSPAS-SARSAT information	1 week	15

P

6.2.9 Singapore

Nanyang Technological University (School of Electrical and Electronics Engineering)

The Nanyang Technological University is one of two institutions providing tertiary education in Singapore. It serves as the national contact point for the Regional Working Group on Satellite Communication Applications and Distance Education. Details of its activities in this field may be obtained by accessing the URL <http://www.ntu.ac.sg>.

Professional Personnel

Primary field of work	Number of staff with the following qualifications			Number of staff with intensive training (>2 consecutive months) during last 5 years
	Ph.D.	MSc	BSc	
Rural telecommunications: education				
Rural telecommunications: health care				
Broadcasting of meteorological information: weather forecasts				
Broadcasting of meteorological information: disaster warnings and relief				
Hardware/software development				
Development of ground terminals (TVROs, VSATs, portable terminals)	17	1		3
Data management/information systems services				

Total number of professional staff: 21   Total number of administrative staff: 2

6.2.10 Sri Lanka

Arthur C. Clarke Centre for Modern Technologies

The Arthur C. Clarke Centre for Modern Technologies is a government organization with the mandate to introduce and accelerate development of space science and technology and related fields in Sri Lanka. As the national focal point for space activities in Sri Lanka, the Centre plays a major role in coordinating space programmes and projects at national, regional and international levels. Its organization chart is shown in figure 6.3.

Professional Personnel

Primary field of work	Number of staff with the following qualifications			Number of staff with intensive training (>2 consecutive months) during last 5 years
	Ph.D.	MSc	BSc	
Rural telecommunications: education				
Rural telecommunications: health care				
Broadcasting of meteorological information: weather forecasts				
Broadcasting of meteorological information: disaster warnings and relief				
Hardware/software development				
Development of ground terminals (TVROs, VSATs, portable terminals)		2		3
Data management/information systems services		2		

Total number of professional staff: 4   Total number of administrative staff: 1

Training Activities (1992-1996)

Course title and description	Training undergone by organization staff		
	Given by (name of institution/organization)	Duration/ frequency	Number of staff taking course
Satellite communications course	United Nations Space Science and Technology Education Centre, Ahmedabad, India	9 months	2

S

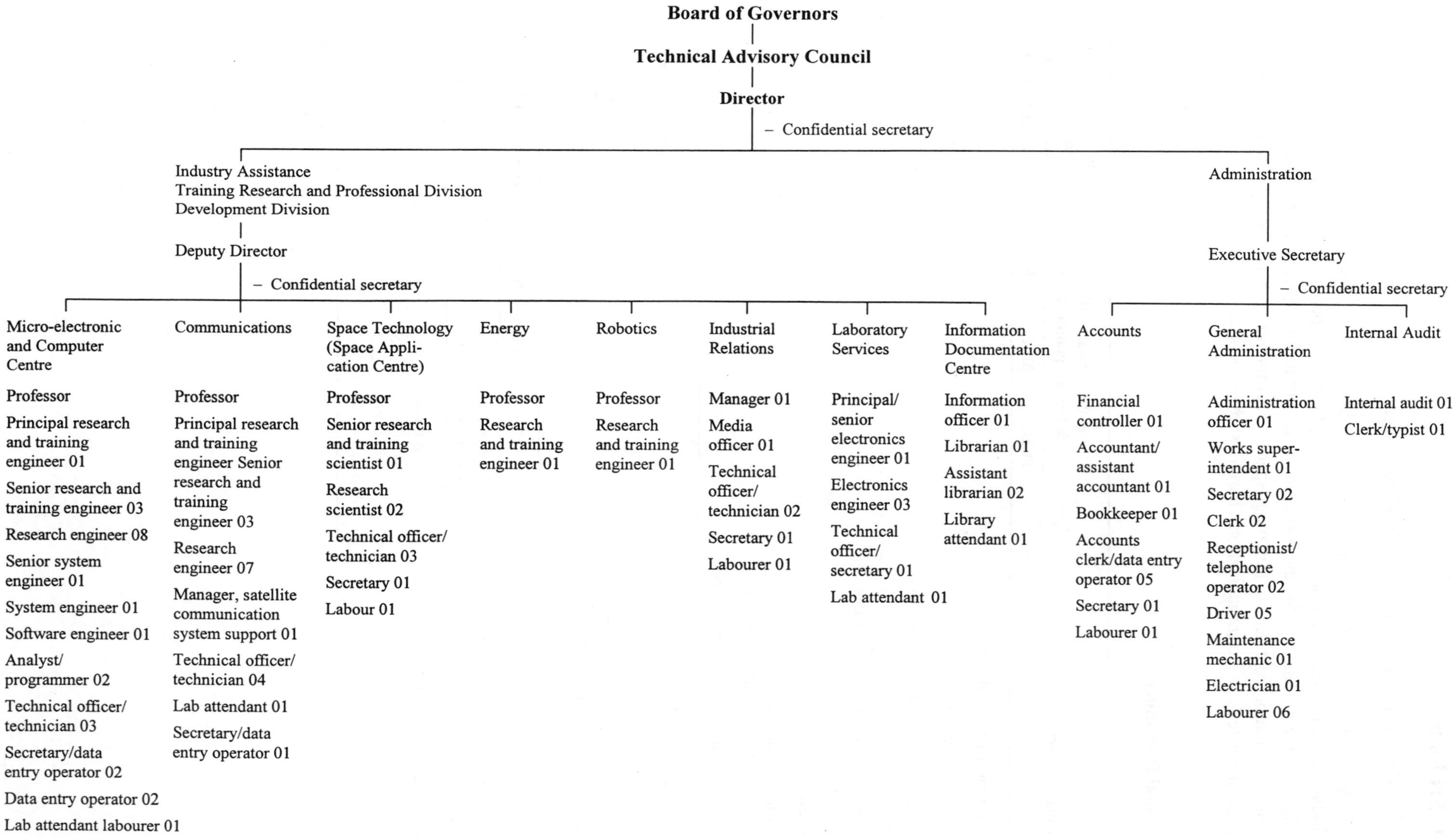

Board of Governors

Technical Advisory Council

Director
— Confidential secretary

Administration
Executive Secretary
— Confidential secretary

Deputy Director
— Confidential secretary

Industry Assistance
Training Research and Professional Development Division

Research and Professional Division

Micro-electronic and Computer Centre	Communications	Space Technology (Space Application Centre)	Energy	Robotics	Industrial Relations	Laboratory Services	Information Documentation Centre
Professor	Professor	Professor	Professor	Professor	Manager 01	Principal/senior electronics engineer 01	Information officer 01
Principal research and training engineer 01	Principal research and training engineer Senior research and training engineer 03	Senior research and training scientist 01	Research and training engineer 01	Research and training engineer 01	Media officer 01	Electronics engineer 03	Librarian 01
Senior research and training engineer 03	Research engineer 07	Research scientist 02			Technical officer/technician 02	Technical officer/secretary 01	Assistant librarian 02
Research engineer 08	Senior system engineer 01	Technical officer/technician 03			Secretary 01	Lab attendant 01	Library attendant 01
Senior system engineer 01	Manager, satellite communication system support 01	Secretary 01			Labourer 01		
System engineer 01	Technical officer/technician 04	Labour 01					
Software engineer 01	Lab attendant 01						
Analyst/programmer 02	Secretary/data entry operator 01						
Technical officer/technician 03							
Lab attendant 01							
Secretary/data entry operator 02							
Data entry operator 02							
Lab attendant labourer 01							

Administration

Accounts	General Administration	Internal Audit
Financial controller 01	Administration officer 01	Internal audit 01
Accountant/assistant accountant 01	Works superintendent 01	Clerk/typist 01
Bookkeeper 01	Secretary 02	
Accounts clerk/data entry operator 05	Clerk 02	
Secretary 01	Receptionist/telephone operator 02	
Labourer 01	Driver 05	
	Maintenance mechanic 01	
	Electrician 01	
	Labourer 06	

Figure 6.3 Organization chart of the Arthur C. Clarke Centre for Modern Technologies

6.2.11 Thailand

Thaicom Distance Education Centre

Thaicom Distance Education Centre is a government organization which serves as the central management and coordination unit for distance education. It has responsibility for formulating policies and plans for conducting research and development, managing overall implementation of the project, coordinating with other government and non-government agencies and the private sector to carry out the satellite distance education programme.

Professional Personnel

Primary field of work	Number of staff with the following qualifications			Number of staff with intensive training (>2 consecutive months) during last 5 years
	Ph.D.	MSc	BSc	
Rural telecommunications: education	1	10	3	
Rural telecommunications: health care				
Broadcasting of meteorological information: weather forecasts				
Broadcasting of meteorological information: disaster warnings and relief				
Hardware/software development				
Development of ground terminals (TVROs, VSATs, portable terminals)				
Data management/information systems services				

Total number of professional staff: 11   Total number of administrative staff: 3

Major Current/Ongoing Projects/Activities

Type of electronic media used	Subject/topic	Co-sponsorship (if any)	Nature and size of intended public
Radio and DTH satellite	All subjects in the basic education curriculum for primary and secondary as well as vocational certificate programmes	Thaicom Foundation	Distance learners (nationwide enrolment is about four million)

Electronic Media Programmes Developed since 1990 and Available to Outside Users

Type (radio, television, computer network, other)	Programme contents (subject and other relevant information)	Year of production	Intended public
Television	Science, technology, mathematics, foreign languages, arts, music, health, environment and culture	1994-1996	Students, NFE learners and general public

T

Documents Published since 1990

Title	Year of publication	Language of publication	Keywords	Frequency
Satellite Distance Education	1995	English	Satellite/distance education, integrated media	
Satellite distance education journal	1995-1996	Thai	Satellite/distance education	Monthly
Satellite distance education training packages	1995	Thai	Satellite/distance education, training	
Satellite distance education booklet	1995	Thai	Satellite/distance education	
Evaluation of satellite distance education	1995-1996	Thai	Satellite/distance education, evaluation	Annual

Training Activities (1992-1996)

Course title and description	Training undergone by organization staff		
	Given by (name of institution/organization)	Duration/ frequency	Number of staff taking course
Computer software applications	Governments and private training institutes	2-4 weeks	5
Distance education management	Algonquin College, Canada	4 weeks	1

Course title and description	Training provided by organization for external users		
	Name of participating agencies/organizations	Duration/ frequency	Number of participants
Satellite education management	NFE regional, provincial and district centres, schools and educational institutions	3 days	4,000 top managers
Utilization of satellite instruction	Schools and NFE learning centres	3 days	12,000 teachers and facilitators

Post and Telegraph Department

The Post and Telegraph Department is a government organization mandated to regulate and monitor domestic satellite communication networks and applications, and to operate and manage domestic satellite communication networks for civil communications. The Department is empowered by Article 3 of the Royal Decree on the Structural Organization of the Post and Telegraph Department to improve both system and technologies in order to present the Department's views, policies, measures and proposed tariff of postal and telecommunication services in the country; represent the government at an administrative level in dealing with the governments and organizations of foreign countries and international organizations in the field of post and telecommunication development; manage the radio frequency spectrum; issue radiocommunication licences; and set and inspect technical specifications and standards of radio communication equipment and accessories.

Professional Personnel

Primary field of work	Number of staff with the following qualifications			Number of staff with intensive training (>2 consecutive months) during last 5 years
	Ph.D.	MSc	BSc	
Rural telecommunications: education			5	
Rural telecommunications: health care				
Broadcasting of meteorological information: weather forecasts				
Broadcasting of meteorological information: disaster warnings and relief				
Hardware/software development		5	25	
Development of ground terminals (TVROs, VSATs, portable terminals)			5	
Data management/information systems services			5	

Total number of professional staff: 15   Total number of administrative staff:

Major Current/Ongoing Projects/Activities

Type of electronic media used	Subject/topic	Co-sponsorship (if any)	Nature and size of intended public
	Development planning		
	Information dissemination		
	Technology development		

Documents Published since 1990

Title	Year of publication	Language of publication	Keywords	Frequency
Annual report		Thai/English		
National communication day		Thai		Annual
Glossary of terms, abbreviations and acronyms (space development)	In preparation	English		

Training Activities (1992-1996)

Course title and description	Training undergone by organization staff		
	Given by (name of institution/organization)	Duration/ frequency	Number of staff taking course
Satellite link design and infrastructure design	Asia-Pacific Satellite Communications Council	1 week	1
TDMA technical training	Comstream A Spar Company, United States	15 days	3
Satellite communications (IBS/IDR)	USTTI	2 weeks	1

Course title and description	Training undergone by organization staff		
	Given by (name of institution/organization)	Duration/ frequency	Number of staff taking course
Satellite communications engineering	MPT, Japan		1
Frequency management	MPT, Japan		1
Telecommunications engineering	JICA		1
Radio communications engineering	Japan	3 months	1
Digital transmission system	Japan		1
Radio telephone operation in fishing ships			930 per year
Operation of synthesizer radio communication			597 per year
Radio telephone operator of ships with international navigation	Government agencies and interested parties		34 per year
Television broadcasting for technicians			27 per year
TDMA technical training	Government agencies	1 week, biannual	15 per year

Communication Authority of Thailand

The Communication Authority of Thailand is a semi-autonomous organization responsible for Thailand's international communications. It has been authorized to provide all telecommunication applications, except the basic domestic telephony, throughout the country. All services such as international telephone, television broadcasting and data communications have been made available by the Communication Authority of Thailand.

Professional Personnel

Primary field of work	Number of staff with the following qualifications			Number of staff with intensive training (>2 consecutive months) during last 5 years
	Ph.D.	MSc	BSc	
Rural telecommunications: education				
Rural telecommunications: health care				
Broadcasting of meteorological information: weather forecasts				
Broadcasting of meteorological information: disaster warnings and relief				
Hardware/software development				
Development of ground terminals (TVROs, VSATs, portable terminals)				
Data management/information systems services				
Systems Engineering and Communications	2	10	40	

Total number of professional staff: 47

Total number of administrative staff: 5

6.2.12 Viet Nam

Department General of Posts and Telecommunications

The Department General of Posts and Telecommunications is the government administrative department in charge of policy-making and regulatory matters on posts and telecommunication and radio frequency management nationwide. It establishes the laws and policies on posts and telecommunication and radio frequency management; issues, within its jurisdiction, regulations on technical and economic standards, networks, services, posts and telecommunication equipment, transponders and radio frequency management; issues and withdraws licences and certificates in the field; and settles and resolves disputes on radio frequency and posts and telecommunication network services.

Professional Personnel

Primary field of work	Number of staff with the following qualifications			Number of staff with intensive training (>2 consecutive months) during last 5 years
	Ph.D.	MSc	BSc	
Rural telecommunications: education	1	2	7	
Rural telecommunications: health care				
Broadcasting of meteorological information: weather forecasts				
Broadcasting of meteorological information: disaster warnings and relief				
Hardware/software development				
Development of ground terminals (TVROs, VSATs, portable terminals)				
Data management/information systems services		60		

Total number of professional staff: 70

Total number of administrative staff: N/A

Major Current/Ongoing Projects/Activities

Type of electronic media used	Subject/topic	Co-sponsorship (if any)	Nature and size of intended public
	Multipurpose community telecentre	ITU, WHO, UNESCO, UNIDO and SIDA	Population in rural areas
	Viet Nam communication satellite (VINASAT)	Government agencies and broadcasting organizations	Entire nation

Training Activities (1992-1996)

Course title and description	Training undergone by organization staff		
	Given by (name of institution/organization)	Duration/ frequency	Number of staff taking course
Digital microwave engineering	Posts and Telecommunications Training Centre No. 1	8 weeks	425
Fibre optics communications	Posts and Telecommunications Training Centre No. 1	12 weeks	143

V

Electronic Media Programmes Developed since 1990 and Available to Outside Users

Type (radio, television, computer network, other)	Programme contents (subject and other relevant information)	Year of production	Intended public
Video tape	Digital microwave	1995-1996	
Video tape	Fibre optics communications	1995-1996	
Video tape	English for telecommunications	1996-1997	

Documents Published since 1990

Title	Year of publication	Language of publication	Keywords	Frequency
Digital microwave engineering	1996	Vietnamese		
Fibre optics communications	1996	Vietnamese		Annual
English for telecommunications	1997	English		

Chapter 7

SPACE SCIENCES AND TECHNOLOGY APPLICATIONS

7.1 Regional overview

7.1.1 Background

The last four decades since the advent of the Space Age in 1957 have witnessed continuous developments in space science and space technology that have had an impact on human lives and activities in countless ways. Over this period, there has been tremendous accumulation of scientific knowledge about the Earth, the near Earth environment, the solar system, outer space and the universe as a whole, while space technology has advanced at an astonishing speed, spurring, in the process, rapid developments in numerous science and engineering disciplines. The applications and derivatives of space technology have correspondingly multiplied at an ever-increasing pace. There is, indeed, no looking back, as one achievement follows another in quick succession.

The variety of services, functions and tasks performed by satellites and other spacecraft reflects the versatility and power of space technology and its increasing involvement in different activities. As is well known, space technology is being routinely used for communications, weather forecasting, atmospheric research, mapping, resource and environmental management, navigation, manufacturing, and scientific research in astronomy, medicine, biology and other fields. It is, in fact, recognized today as an important enabling technology for socio-economic development. ESCAP has taken due cognizance of the role of space science and technology as enabling tools in socio-economic progress. In order to promote the potential uses of space science and technology, ESCAP has established, under the framework of RESAP, the Regional Working Group on Space Sciences and Technology Applications to deal with selected fields that are of interest to countries participating in the Working Group.

7.1.2 Status of space science and technology applications in the region

As the Asian and Pacific region consists mostly of developing countries, the national space programmes in the region are, in general, focused on those fields or aspects of space science and technology that are relevant to their needs and have immediate and direct applications.

The four spacefaring countries in the region – China, India, Japan and the Russian Federation – have comprehensive space programmes covering many fields. They have strong space science programmes carried out in research institutions and universities encompassing various fields such as middle and upper neutral atmosphere research, climate change studies, ionospheric physics and radio wave propagation, magnetospheric physics, plasma physics, solar physics, cosmic rays, microgravity research, optical, infra-red, radio and X-ray astronomy, planetary science and astrophysics, with good parallel programmes in the physical, Earth and life sciences. The science component provides a firm basis for indigenous space technology development in these countries, which have made commendable progress in many fields. These countries are engaged in developing their capabilities in a wide range of technologies in various R and D organizations, backed by a sound and expanding industrial base. They have designed, developed and fabricated operational meteorological, remote sensing and communication satellites as well as scientific and experimental satellites, which have, by and large, performed satisfactorily in polar or inclined and geostationary orbits; they operate sophisticated ground receiving/TT and C stations for these satellites. They also develop and launch sounding rockets and balloons for upper atmosphere research. The four countries have been able to develop their own launch vehicles, which have been largely successful (barring some failures) in placing in orbit their satellites as well as, in some cases, satellites from other countries, hence the appellation "spacefaring" used of them. Japan and the Russian Federation have also launched various spacecraft for scientific investigations of, for instance, the moon, comets, magnetosphere, aurora and X-ray stars. The Russian Federation has also sent probes to other planets and operates Mir, the only space station in orbit since 1986. Japan and the Russian Federation will be major partners with the United States, Europe and Canada in building and operating the international permanent space station due to start functioning in 2002.

The space science and technology programmes of countries placed in the second category (see the regional overview in chapter 3) are, in general, relatively smaller than those of the spacefaring nations. Nevertheless, space science programmes in most countries in this category include some or all of the following fields: middle and upper neutral atmosphere research, climate change studies, ionospheric physics and radio wave propagation, solar-terrestrial physics, astronomy and astrophysics. The space technology component in these countries usually covers operation of ground stations for remote sensing, communication and meteorological satellites, development of remote sensors, development and launching of sounding rockets and use of balloons for upper atmosphere research, and development of small satellites. Australia, a developed country, has a wide-ranging space programme and possesses extensive facilities and expertise in both science and technology. It also plans to develop satellite launch capabilities for commercial launches. Other countries that are involved in several space science and space technology fields, as mentioned above, and possess reasonably well-developed capabilities in some fields are Azerbaijan, Indonesia, the Islamic Republic of Iran, Pakistan, the Philippines, the Republic of Korea, Singapore and Thailand.

Some countries in the region plan to develop their indigenous space technology capabilities. Australia, China, Indonesia, Malaysia, Pakistan, the Philippines, the Republic of Korea, Singapore and Thailand are engaged in the development of small/micro LEO satellites (sometimes with technical support from agencies in developed countries) for various applications such as communications, remote sensing of resources, environmental monitoring, disaster monitoring or other scientific experiments. Several regional countries, besides the spacefaring ones, have their own national satellite communication systems. Although the communication satellites are often developed and launched by foreign companies, engineers and scientists from the concerned regional countries are sometimes associated in the development work with the foreign companies, as part of package deals involving, *inter alia,* technology transfer, and are thus able to acquire some know-how in the field, which can be useful when efforts are made to enhance indigenous capabilities in these countries. However, those countries which have been placed in the third and fourth categories in the regional overview in chapter 3 have little or no involvement in fields other than remote sensing and GIS, satellite meteorology and satellite communications.

Keeping in view the present status of space science and technology, as well as the main interests of regional countries, the Regional Working Group on Space Sciences and Technology Applications, at its second meeting held in Singapore in February 1997, identified five priority areas, namely (a) developing a simple common payload for small satellites, (b) application/sharing of small satellite data, (c) low-cost ground stations for small satellites, (d) developing electronic media for information exchange on space science and technology and (e) study of a joint regional satellite programme for its medium-term plan. It is hoped that the collaborative activities and projects to be undertaken under the medium-term plan will enhance the capabilities of individual participating countries and the region as a whole in these priority areas.

7.1.3 Conclusion

As is evident from the foregoing, the vast majority of developing regional countries do not have, in general, very well-established space science and technology programmes in fields other than remote sensing, GIS, satellite-based positioning, satellite meteorology, disaster monitoring, satellite communications and distance education. Apart from the four spacefaring countries, which are engaged in comprehensive space programmes, there are about 10-12 other countries in the region that are pursuing relatively wide-ranging space programmes that include various fields of space science and space technology besides the three broad fields mentioned above. While the number of regional countries that are becoming involved in a wider range of space science and space technology activities is increasing with time, this increase is still quite gradual for obvious reasons. Nevertheless, it is hoped that space science and technology applications will increasingly contribute to sustainable development activities in the region.

Table 7.1 National contact points of the Regional Working Group on Space Sciences and Technology Applications

Country	Name	Agency
Australia	Mr David Jauncey	Australia Telescope National Facility
Azerbaijan	–	–
Bangladesh	Mr D.A. Quader	Space Research and Remote Sensing Organization
China	Mr Sun Huixian	Centre for Space Sciences and Technology Applications
Fiji	Mr Les Allison Mr Rupert Anise	SOPAC Ministry of Agriculture, Fisheries, Forest and ALTA
India	–	–
Indonesia	Mr Soewarto Hardhienata	National Institute of Aeronautics and Space
Islamic Republic of Iran	Mr Mohammad Hasan Entezari	Data Communication of Iran
Japan	Mr Akira Noie	Science and Technology Agency
Malaysia	Professor Mazlan Othman	Planetarium Negara
Mongolia	Mr M. Ganzorig	Institute of Information, Academy of Sciences
Myanmar	–	–
Nepal	Mr Keshar M. Bajracharya	Royal Nepal Academy of Science and Technology
Pakistan	Mr M. Ishaq Mirza	Space and Upper Atmosphere Research Commission
Philippines	Mr Victorio Ochave	Advanced Science and Technology Institute
Republic of Korea	Professor Soon Dal Choi	Satellite Technology Research Centre
Russian Federation	–	–
Singapore	–	–
Sri Lanka	Director	Arthur C. Clark Centre for Modern Technologies
Thailand	Mr Suthi Aksornkitti	King Mongkut Institute of Technology
Vanuatu	–	–
Viet Nam	Mr Nguyen Quang Thinh	National Centre for Science and Technology

7.2 Country inputs

7.2.1 Australia

CSIRO Office of Space Science and Applications

The Commonwealth Scientific and Industrial Research Organization (CSIRO) is one of Australia's largest and oldest scientific research bodies. Its tasks are prescribed by the Science and Industry Research Act 1949, and include carrying out research and development which further the interests of the Australian community, contributes to national objectives, and benefits Australian industry.

Many of the responsibilities of CSIRO require access to space technology: for exploring the universe; monitoring and trying to understand global change; testing new procedures in mineral exploration; and developing goods and services of increasing economic importance to Australia. Space is a unique vantage point for observation and communication systems which are important for scientific research, for services, and to the global information economy.

Planning the use of space technologies in CSIRO, contributing to international cooperation in sharing global space assets, and managing space projects are tasks carried out by the CSIRO Office of Space Science and Applications (COSSA), established in 1984. About half of the twenty-plus divisions of CSIRO carry out space-related research and services. Some of the areas in which these are carried out are given below.

Radioastronomy

Many cosmic objects, from individual stars to whole galaxies, emit radio waves. These signals tell us about their behaviour and structure. CSIRO has been active in radio astronomy for over 50 years. The CSIRO Australia Telescope National Facility (ATNF) is one of the world's premier radioastronomy laboratories. The Australia Telescope, a bicentennial project, is an outstanding example of Australian scientific endeavour, engineering, project management and construction.

The Facility includes (a) Parkes Observatory 64-metre dish, (b) Australia Telescope Compact Array, Narrabri, and (c) Mopra Observatory, Coonabarabran.

Communications

Engineering and research work in space communication is carried out in CSIRO Telecommunications and Industrial Physics. CTIP has designed correlator chips and other very large-scale integrated circuits now in commercial use for many applications, including radar signal processing.

CSIRO designed and fabricated the feed horn for the Tasmanian Earth Resources Satellite Station in Hobart, and carries out electromagnetic design for communication antennas. As well, it assists industry to commercialize this technology. CSIRO won a contract from Hughes Satellites for design work on the HS 601 C and the Optus satellite series.

The planned FedSat microsatellite, a Centenary of Federation project which will see a low-cost Australian space mission starting in 2001, is to carry engineering payloads and scientific experiments including satellite to satellite communications, location-finding transmitters, a magnetometer and a Ka-band beacon. The payloads will be developed through a proposed cooperative research centre for satellite systems.

Spacecraft tracking

The ATNF has supported critical spacecraft tracking operations for NASA and the European Space Agency, including the Voyager encounters with Neptune and Uranus and the Giotto probe to Halley's Comet. CTIP oversees an agreement between Australia and the United States under which NASA spacecraft are tracked by the Tidbinbilla Deep Space Communication Complex. The facility, which is one of Canberra's most popular tourist destinations, is also used by CSIRO for astronomical research in cooperation with NASA. It is operated by British Aerospace Australia under contract, and employs about 130 skilled staff.

Instrumentation

Several CSIRO divisions, including Manufacturing Science and Technology, Exploration and Mining; Atmospheric Research; and Telecommunications and Industrial Physics have demonstrated abilities in space or airborne optics and measurement instrumentation.

This work includes the award-winning tunable MIRRACOLAS CO_2 laser system for mineral mapping; completed Phase A and B studies (with Vipac Engineering) on a future spaceborne Atmospheric Pressure Sensor; testing of a prototype airborne infra-red sensor for detecting volcanic ash clouds; design of the L-band low noise amplifier for the Russian-led RADIOASTRON space radioastronomy mission; and airborne hyper-spectral infra-red scanners for mineral exploration and environmental monitoring. CSIRO scientists carried out fundamental design work contributing to the Along Track Scanning Radiometers currently flying on the European Remote Sensing satellites.

CSIRO is currently participating, with industry and government parties, in plans for the ARIES-1 spaceborne hyper-spectral sensing system for mineral exploration and environmental applications.

Propulsion and guidance systems

CSIRO is a patron of the Australian Space Research Institute, a non-profit body which undertakes, with industry support, university-based scientific and engineering research on small rockets and satellites.

Earth observation

CSIRO activities in airborne and space-based remote sensing are intensive and geographically dispersed. About 70 professionals in ten or so divisions take part in strategic research in remote sensing science, or in the reception, management, processing, evaluation and application of Earth observation data and products. The focus of scientific research in this field is the CSIRO Earth Observation Centre, a programme hosted by the CSIRO Office of Space Science and Applications.

International cooperation

The exploration and use of space now involves more nations than ever before, with many countries, especially in the Asian and Pacific region, recently starting ambitious space projects. CSIRO cooperates with many of these international space agencies and research agencies.

Through the Australia Telescope, CSIRO is now taking part in astronomical research relying on observations from a space-based radio astronomy mission, VSOP/Muses-B or HALCA, launched by the Institute of Space and Astronautical Science from Japan on 11 February 1997.

CSIRO scientists are investigators for many NASA space missions, using measurements taken by unmanned satellites to observe both space and the Earth. They also use data and results from the extensive space programmes of NASDA, the National Space Development Agency of Japan; and from the European Space Agency. CSIRO has also conducted space experiments with the French space agency, CNES; with the Indian Space Research Organization; with the Department of the Environment in the United Kingdom, the Russian Institute for Space Research, and with the Chinese Academy of Sciences, amongst others.

International coordination is important for reducing the cost and increasing the effectiveness of space missions. CSIRO takes an active role in regional space affairs, through the technical studies and projects of ESCAP. The organization also represents Australia at the Space Agency Forum and at the International Astronautical Federation, the 1998 Congress of which is in Melbourne.

COSSA is the Australian representative on the international Committee on Earth Observation Satellites, and chaired CEOS in 1995-1996.

Source: COSSA Web site, <www.cossa.csiro.au>.

7.2.2 Azerbaijan

Azerbaijan National Space Agency

The Azerbaijan National Space Agency (ANASA) is a government organization whose mandate is specified in a decree issued by the President of Azerbaijan.

Professional Personnel

Primary field of work	Number of staff with the following qualifications			Number of staff with intensive training (>2 consecutive months) during last 5 years
	Ph.D.	MSc	BSc	
Fabrication and testing of satellites				
Fabrication and testing of satellite launchers				
Fabrication and testing of sounding rockets				
Construction of launch facilities				
Development of telemetry equipment				
Development of in-flight data acquisition and analysis facilities	8	2	70	
Development of inexpensive data collection platforms				
Development and launching of equatorial-orbiting satellites				
Development and launching of small satellites				
Development of scientific satellites				
Participation in joint space science experiments	1	2	20	

Total number of professional staff: 103   Total number of administrative staff: N/A

Field of Activity Being Pursued

Field of activity	Description of major ongoing activities	Description of interaction with other developing contries and/or efforts towards addressing their needs
Fabrication and testing of satellites		
Fabrication and testing of satellite launchers		
Fabrication and testing of sounding rockets		
Construction of launch facilities		
Development of telemetry equipment	Development of autonomous ground measure stations	
Development of in-flight data acquisition and analysis facilities	Development of onboard measure-informative system	

Field of activity	Description of major ongoing activities	Description of interaction with other developing contries and/or efforts towards addressing their needs
Development of inexpensive data collection platforms		
Development and launching of equatorial-orbiting satellites		
Development and launching of small satellites	Development of X-ray spectrometer	
Development of scientific satellites		
Participation in joint space science experiments		

Training Activities (1992-1996)

Course title and description	Training provided by organization for external users		
	Name of participating agencies/organizations	Duration/ frequency	Number of participants
Earth study from space; aerospace device development; measure-information systems; automatic systems for information processing and control	Azerbaijan State Oil Academy and State Technology University	Annual	95

Documents Published since 1990

Title	Year of publication	Language of publication	Keywords	Frequency
Communication of science-industrial association of space research, Baku	1990	Russian	Remote sensing, sensors, processors	
Space science and technology symposium, Turkey	1993	English	Monitoring, ecology, aerospace information processing	

.2.3   China

General Establishment of Space Science and Applications, Chinese Academy of Sciences

The General Establishment of Space Science and Applications of the Chinese Academy f Sciences is a semi-autonomous institution coordinating all space activities of the 49 institutes nder the Academy.   It also promotes and coordinates cooperative activities with other national rganizations as˙well as other countries.   Its organization chart is shown in figure 7.1.

rofessional Personnel

rimary field of work	Number of staff with the following qualifications			Number of staff with intensive training (>2 consecutive months) during last 5 years
	Ph.D.	MSc	BSc	
abrication and testing of satellites	20	20	10	30
abrication and testing of satellite unchers				
abrication and testing of sounding ockets	10	10		
onstruction of launch facilities				
evelopment of telemetry equipment	5	5		
evelopment of in-flight data acquisition ıd analysis facilities	10	20	10	
evelopment of inexpensive data ollection platforms	5	5		
evelopment and launching f equatorial-orbiting satellites				
evelopment and launching f small satellites	6			
evelopment of scientific satellites	20	10	10	
articipation in joint space science xperiments	20	20	10	

otal number of professional staff: 103   Total number of administrative staff: 160

ield of Activity Being Pursued

ield of activity	Description of major ongoing activities	Description of interaction with other developing contries and/or efforts towards addressing their needs
abrication and testing of satellites		
abrication and testing of satellite unchers		
abrication and testing of sounding ockets	ZN-1, ZN-3 sounding rockets, tracking	Tracking responder being exported to Indonesia
onstruction of launch facilities		
evelopment of telemetry equipment	S-band (being developed) and UHF (developed)	

China

Organization

Chairman | 169

Pers–3
1 x Chairman
1 x Stenographer (PA)
1 x MISS

| 38

6
irector
enographer (PA)
1SS
killed Worker/Laboratory
ttendant

Vater Resources Division

6
hief Scientific Officer
incipal Scientific Officer/
incipal Engineer
enior Scientific Officer/
nior Engineer
cientific Officer/
ssistant Engineer
ssistant Scientific Officer/
nior Engineer
echnician Grade II

Fisheries Division

5
incipal Scientific Officer/
incipal Engineer
nior Scientific Officer/
nior Engineer
cientific Officer/
ssistant Engineer
ssistant Scientific Officer/
nior Engineer
echnician Grade I/
ientific Assistant

Oceanography Division

5
incipal Scientific Officer/
incipal Engineer
nior Scientific Officer/
nior Engineer
cientific Officer/
ssistant Engineer
ssistant Scientific Officer/
nior Engineer
echnician Grade I/Scientific Assistant

ion

Technology | 50

Pers–7
1 x Director
1 x Stenographer (PA)
1 x MISS
4 x Skilled Worker/Laboratory
   Attendant

Ground Station

Pers–12
1 x Chief Scientific Officer
1 x Principal Scientific Officer/
   Principal Engineer
1 x Senior Scientific Officer/
   Senior Engineer
3 x Scientific Officer/
   Assistant Engineer
1 x Assistant Scientific Officer/
   Junior Engineer
1 x Senior Technician/
   Senior Scientific Assistant
1 x Mechanic/Plumber/
   Refrigeration Mech.
1 x Laboratory Assistant
1 x Technician Grade I/
   Scientific Assistant
1 x Technician Grade II

Photographic Division

Pers–6
1 x Senior Scientific Officer/
   Senior Engineer
2 x Scientific Officer/
   Assistant Engineer
1 x Assistant Scientific Officer/
   Junior Engineer
1 x Technician Grade I/
   Scientific Assistant
1 x Technician Grade II

Rocket Technology Development Division

Pers–3
1 x Senior Scientific Officer/Senior Engineer
1 x Scientific Officer/Assistant Engineer
1 x Assistant Scientific Officer/Junior Engineer

Instrumentation and Data Processing Division

Pers–13
1 x Chief Scientific Officer
1 x Principal Scientific
   Officer/Principal Engineer
1 x Senior Scientific Officer/
   Senior Engineer
2 x Scientific Officer/
   Assistant Engineer
1 x Assistant Scientific Officer/
   Junior Engineer
2 x Senior Technician/
   Senior Scientific Assistant
1 x Mechanic/Plumber/
   Refrigeration Mech.
2 x Laboratory Assistant
1 x Technician Grade I/
   Scientific Assistant
1 x Technician Grade II

Cartography Division

Pers–9
1 x Principal Scientific Officer/
   Principal Engineer
1 x Senior Scientific Officer/
   Senior Engineer
2 x Scientific Officer/
   Assistant Engineer
1 x Assistant Scientific Officer/
   Junior Engineer Assistant
   Cartographer
3 x Draughtsman
1 x Technician Grade II

Research

Pers–5
1 x Director
1 x Stenographer (PA)
1 x MISS
4 x Skilled Worker

Atmospheric Division

Pers–5
1 x Principal Scientific Officer/
   Principal Engineer
1 x Senior Scientific Officer/
   Senior Engineer
1 x Scientific Officer/
   Assistant Engineer
1 x Assistant Scientific Officer/
   Junior Engineer
1 x Senior Technician/
   Senior Scientific Assistant

Ocean Physics Division

Pers–3
1 x Senior Scientific Officer/
   Senior Engineer
1 x Scientific Officer/
   Assistant Engineer
1 x Assistant Scientific Officer/
   Junior Engineer

Agro-Hydro-Meteorology Division

Pers–4
1 x Chief Scientific Officer
1 x Senior Scientific Officer/
   Senior Engineer
1 x Scientific Officer/
   Assistant Engineer
1 x Assistant Scientific Officer/
   Junior Engineer

Space Physics and Rocket Dynamics Division

Pers–3
1 x Senior Scientific Officer/
   Senior Engineer
1 x Scientific Officer/Assistant
   Engineer
1 x Assistant Scientific Officer/
   Junior Engineer

---

7. To set up both resource and weather satellite ground stations.
8. To develop instrumentation facilities for multidisciplinary applications of remote sensing data.
9. To carry out research and development activities in the fields of agriculture, water resources, foresty, fisheries, meteorology geology, oceanography,
10. To establish regional and international cooperation and collaboration in the relevant fields.
11. To provide relevant information to the Government of Bangladesh in formulating national and international policy issues concerning space science and remote sensing activities.
12. Any other duty assigned by the Authority.

Chinese Academy of Sciences

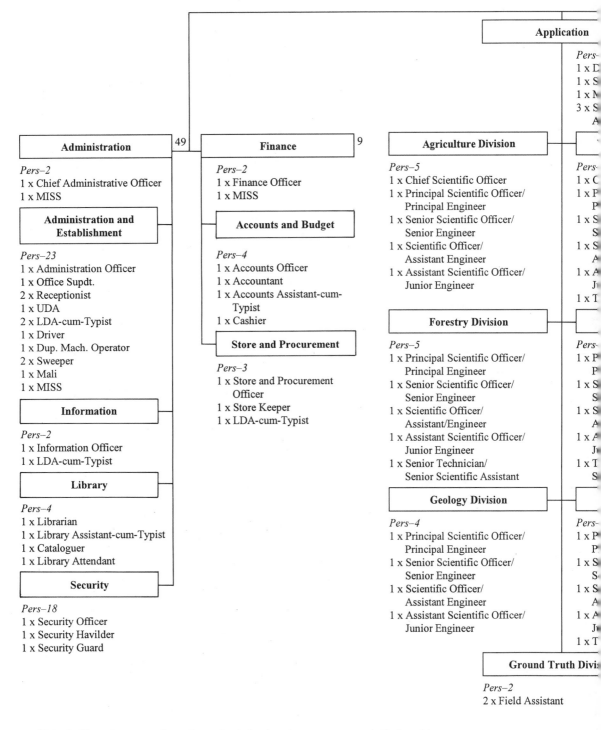

Administration | 49

Pers–2
1 x Chief Administrative Officer
1 x MISS

Administration and Establishment

Pers–23
1 x Administration Officer
1 x Office Supdt.
2 x Receptionist
1 x UDA
2 x LDA-cum-Typist
1 x Driver
1 x Dup. Mach. Operator
2 x Sweeper
1 x Mali
1 x MISS

Information

Pers–2
1 x Information Officer
1 x LDA-cum-Typist

Library

Pers–4
1 x Librarian
1 x Library Assistant-cum-Typist
1 x Cataloguer
1 x Library Attendant

Security

Pers–18
1 x Security Officer
1 x Security Havilder
1 x Security Guard

Finance | 9

Pers–2
1 x Finance Officer
1 x MISS

Accounts and Budget

Pers–4
1 x Accounts Officer
1 x Accountant
1 x Accounts Assistant-cum-Typist
1 x Cashier

Store and Procurement

Pers–3
1 x Store and Procurement Officer
1 x Store Keeper
1 x LDA-cum-Typist

Agriculture Division

Pers–5
1 x Chief Scientific Officer
1 x Principal Scientific Officer/ Principal Engineer
1 x Senior Scientific Officer/ Senior Engineer
1 x Scientific Officer/ Assistant Engineer
1 x Assistant Scientific Officer/ Junior Engineer

Forestry Division

Pers–5
1 x Principal Scientific Officer/ Principal Engineer
1 x Senior Scientific Officer/ Senior Engineer
1 x Scientific Officer/ Assistant/Engineer
1 x Assistant Scientific Officer/ Junior Engineer
1 x Senior Technician/ Senior Scientific Assistant

Geology Division

Pers–4
1 x Principal Scientific Officer/ Principal Engineer
1 x Senior Scientific Officer/ Senior Engineer
1 x Scientific Officer/ Assistant Engineer
1 x Assistant Scientific Officer/ Junior Engineer

Ground Truth Divi

Pers–2
2 x Field Assistant

Note: 1. To apply space and remote sensing technology to resource survey in the launching.
2. To monitor natural hazards in the region.
3. To collect data on renewable and non-renewable resources of the country.
4. To evolve efficient methods for surveying identifying, classifying and monitoring natural resources of the country.
5. To collect agro-hydro-meteorological data daily on real-time basis through several data collecting platforms (DCP) in the country via satellite.
6. To set up a date acquisition, processing and disseminating system for use by different organizations in the country.

Figure 7.1 Organization chart of the General Establishment of Space Science and Applications

Field of activity	Description of major ongoing activities	Description of interaction with other developing contries and/or efforts towards addressing their needs
Development of in-flight data acquisition and analysis facilities	Ground application centre, especially for scientific satellites	
Development of inexpensive data collection platforms		
Development and launching of equatorial-orbiting satellites		
Development and launching of small satellites	In Phase A	
Development of scientific satellites	SJ-5 in Phase B	
Participation in joint space science experiments	Cluster (with ESA), AMS (with United States) and microgravity experiments (with Japan)	

Documents Published since 1990

Title	Year of publication	Language of publication	Keywords	Frequency
Space science journal	Since 1981	Chinese		Quarterly
National report on space research submitted to the Committee on Space Research (COSPAR)	1994, 1996	English		Biennial
ISY in China	1990-1992	English		2-3 per year

7.2.4 Indonesia

National Institute of Aeronautics and Space

The National Institute of Aeronautics and Space (LAPAN) is a government organization serving as the R and D agency for aeronautics and space and as the secretariat of the National Council for Aeronautics and Space of Indonesia (DEPANRI). It also functions as the national focal point for RESAP.

Professional Personnel

Primary field of work	Number of staff with the following qualifications			Number of staff with intensive training (>2 consecutive months) during last 5 years
	Ph.D.	MSc	BSc	
Fabrication and testing of satellites				
Fabrication and testing of satellite launchers	4	24	34	
Fabrication and testing of sounding rockets	4	22	28	
Construction of launch facilities	4	8	22	
Development of telemetry equipment	2	6	16	2 by DLR, Germany
Development of in-flight data acquisition and analysis facilities	2	6	16	2 by DLR, Germany
Development of inexpensive data collection platforms	2	6	16	
Development and launching of equatorial-orbiting satellites	4	12	16	1 by ISRO, India
Development and launching of small satellites	4	8	18	
Development of scientific satellites	4	8	22	
Participation in joint space science experiments	1	4	4	

Total number of professional staff: 301   Total number of administrative staff: 77

Field of Activity Being Pursued

Field of activity	Description of major ongoing activities	Description of interaction with other developing contries and/or efforts towards addressing their needs
Fabrication and testing of satellites		
Fabrication and testing of satellite launchers	Planned satellite launchers: 1997 (30 kg/60 km) and 2003 (50 kg/200 km)	
Fabrication and testing of sounding rockets	R and D activities	
Construction of launch facilities	Embarking upon establishment of space port being considered	

Field of activity	Description of major ongoing activities	Description of interaction with other developing contries and/or efforts towards addressing their needs
Development of telemetry equipment	R and D activities	
Development of in-flight data acquisition and analysis facilities	R and D activities	
Development of inexpensive data collection platforms	R and D activities	
Development and launching of equatorial-orbiting satellites	R and D activities	
Development and launching of small satellites	R and D activities	
Development of scientific satellites	R and D activities	
Participation in joint space science experiments	R and D activities	

Training Activities (1992-1996)

Course title and description	Training undergone by organization staff		
	Given by (name of institution/agency)	Duration/ frequency	Number of staff taking course
Telemetry and telecommand	DLR, Germany	6 months	2
Satellite payloads	DLR, Germany	3 months	1
Satellite technology	ISRO, India	1 year	1
Data communications	DLR, Germany	3 months	3

Documents Published since 1990

Title	Year of publication	Language of publication	Keywords	Frequency
Application of space technology for Indonesian development programme	1994	English	Communications, remote sensing, research	Quarterly
Papers on the RXX-150-LPN-A-DUI and RX-420-LPN rockets				
Papers on planetary nebulae				

7.2.5 Japan

Institute of Space and Astronautical Science

The Institute of Space and Astronautical Science (ISAS) is a national inter-university institute and a central core for space science in Japan. It is one of two major agencies in Japan engaged in space applications. The institute has its own launch vehicle. Its organization chart is shown in figure 7.2.

Professional Personnel

Primary field of work	Number of staff with the following qualifications			Number of staff with intensive training (>2 consecutive months) during last 5 years
	Ph.D.	MSc	BSc	
Fabrication and testing of satellites	62		35	
Fabrication and testing of satellite launchers	78		43	
Fabrication and testing of sounding rockets	78		43	
Construction of launch facilities			9	
Development of telemetry equipment	39		22	
Development of in-flight data acquisition and analysis facilities	41		23	
Development of inexpensive data collection platforms				
Development and launching of equatorial-orbiting satellites				
Development and launching of small satellites	41		25	
Development of scientific satellites	62		35	
Participation in joint space science experiments	62		35	

Total number of professional staff: 227   Total number of administrative staff: 73

Field of Activity Being Pursued

Field of activity	Description of major ongoing activities	Description of interaction with other developing contries and/or efforts towards addressing their needs
Fabrication and testing of satellites	Satellites for astrophysics, solar-terrestrial science and planetary exploration	None
Fabrication and testing of satellite launchers	M-V launch vehicle with solid propellant	None
Fabrication and testing of sounding rockets	S-520, S-310, MT-135 sounding rockets	Cooperation with China, India and Indonesia in the past
Construction of launch facilities	Kagoshima Space Centre (reconstruction is under investigation)	None

304

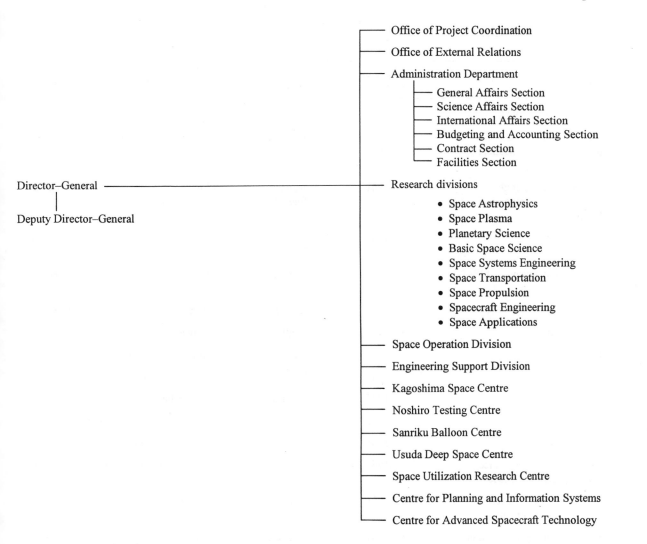

Figure 7.2 **Figure 7.2 Organization chart of the Institute of Space and Astronautical Science**

Field of activity	Description of major ongoing activities	Description of interaction with other developing contries and/or efforts towards addressing their needs
Development of telemetry equipment	Rocket telemetry and satellite telemetry	None
Development of in-flight data acquisition and analysis facilities	Antennas with diameters 64 m, 20 m, 10 m for satellites and 3.8 m and 7 m for rockets; two tracking stations	None
Development of inexpensive data collection platforms	None	None
Development and launching of equatorial-orbiting satellites	None	None
Development and launching of small satellites	cf. ISAS satellite and spacecraft	None
Development of scientific satellites	cf. ISAS in near future	None
Participation in joint space science experiments	cf. International cooperation in scientific satellite programmes	Cooperation with India, China and Indonesia in the past

Documents Published since 1990

Title	Year of publication	Language of publication	Readership	Frequency
House journal of ISAS	1982	Japanese	Members of ISAS	Monthly
Outline of the Institute of Space and Astronautical Science	1981/ 1983	Japanese/ English	Research and public	Annual
Outline of the Institute of Space and Astronautical Science	1981	Japanese/ English	Public	Irregular
Project leaflet	1981	Japanese/ English	Public	Irregular
Outline of other annex of ISAS	1981	Japanese	Public	Every 2 years
ISAS report	1981	Japanese	Researchers and public	Irregular
ISAS report	1981	English	Researchers and public	Irregular
ISAS report, special	1981	Japanese	Researchers and public	Irregular
ISAS report, special	1983	English	Researchers and public	Irregular
ISAS research note	1975	English	Researchers	Irregular
ISAS annual report	1982	Japanese	Researchers and public	Annual
ISAS news	1981	Japanese	Researchers and public	Monthly
Research report on science and engineering of space	1961	Japanese	Researchers and public	Irregular
Research report on science and engineering of balloons	1981	Japanese	Researchers and public	Irregular
Summary on experimentation of sounding rockets	1961	Japanese	Members of ISAS	Annual

Electrotechnical Laboratory, Ministry of International Trade and Industry

The Electrotechnical Laboratory (ETC) of the Ministry of International Trade and Industry is one of the largest national research institutes in Japan. Since its foundation in 1891 as a testing laboratory on electricity, the Laboratory has been undertaking research on electrical, electronic and information technologies. The Laboratory encourages basic research and development in several fields including electronics, information processing, energy technology and standards and measurements. It also houses a space technology section which undertakes research on advanced space technologies including space energy, space robotics, Earth observation sensors and microgravity materials processes.

Professional Personnel

Primary field of work	Number of staff with the following qualifications			Number of staff with intensive training (>2 consecutive months) during last 5 years
	Ph.D.	MSc	BSc	
Fabrication and testing of satellites	3	2	1	
Fabrication and testing of satellite launchers				
Fabrication and testing of sounding rockets				
Construction of launch facilities				
Development of telemetry equipment				
Development of in-flight data acquisition and analysis facilities				
Development of inexpensive data collection platforms				
Development and launching of equatorial-orbiting satellites				
Development and launching of small satellites				
Development of scientific satellites				
Participation in joint space science experiments	3	1	2	

Total number of professional staff: 15   Total number of administrative staff: 670

Field of Activity Being Pursued

Field of activity	Description of major ongoing activities	Description of interaction with other developing contries and/or efforts towards addressing their needs
Fabrication and testing of satellites	Supporting satellite development projects of MITI. Precise telerobotic system on board the ETS-7	
Fabrication and testing of satellite launchers		
Fabrication and testing of sounding rockets		

Field of activity	Description of major ongoing activities	Description of interaction with other developing contries and/or efforts towards addressing their needs
Construction of launch facilities		
Development of telemetry equipment		
Development of in-flight data acquisition and analysis facilities		
Development of inexpensive data collection platforms		
Development and launching of equatorial-orbiting satellites		
Development and launching of small satellites		
Development of scientific satellites		
Participation in joint space science experiments	Space station microgravity experiments on space conductive thin film	

Documents Published since 1990

Title	Year of publication	Language of publication	Keywords	Frequency
Bulletin of the Electrotechnical Laboratory		Japanese		
ETL News		Japanese		
Homepage of the Space Technology Section, ETL		English	<http://www.etl.go.jp/ organization/space/ welcome.html>	
Space robotics research at ETL		English	<http://www.etl.go.jp/ organization/space/ robotics.html>	

7.2.6 Mongolia

National Remote Sensing Centre

The National Remote Sensing Centre of Mongolia is a government organization mandated to coordinate remote sensing and GIS activities in the country, to develop a natural resource management system and to receive and process remotely sensed data. It serves as the national focal point for RESAP.

Professional Personnel

Primary field of work	Number of staff with the following qualifications			Number of staff with intensive training (>2 consecutive months) during last 5 years
	Ph.D.	MSc	BSc	
Fabrication and testing of satellites				
Fabrication and testing of satellite launchers				
Fabrication and testing of sounding rockets				
Construction of launch facilities				
Development of telemetry equipment				
Development of in-flight data acquisition and analysis facilities				
Development of inexpensive data collection platforms				
Development and launching of equatorial-orbiting satellites				
Development and launching of small satellites				
Development of scientific satellites	5	7	10	
Participation in joint space science experiments	4	6	15	

Total number of professional staff: 35   Total number of administrative staff: 10

M

Field of Activity Being Pursued

Field of activity	Description of major ongoing activities	Description of interaction with other developing contries and/or efforts towards addressing their needs
Fabrication and testing of satellites		
Fabrication and testing of satellite launchers		
Fabrication and testing of sounding rockets		
Construction of launch facilities		
Development of telemetry equipment		
Development of in-flight data acquisition and analysis facilities		
Development of inexpensive data collection platforms		
Development and launching of equatorial-orbiting satellites		
Development and launching of small satellites		
Development of scientific satellites	Continuing activity	Interaction with Japan and NASA
Participation in joint space science experiments	Continuing activity	Interaction with Japan and NASA

7.2.7 Pakistan

Space and Upper Atmosphere Research Commission

The Space and Upper Atmosphere Research Commission (SUPARCO) is an autonomous commission of the federal government. As the national space agency, SUPARCO is responsible for the implementation of the national space programme, which is essentially aimed at furthering research in space science and related fields, enhancing capabilities in space technology and promoting the applications of space science and technology for the socio-economic development of the country.

Professional Personnel

Primary field of work	Number of staff with the following qualifications			Number of staff with intensive training (>2 consecutive months) during last 5 years
	Ph.D.	MSc	BSc	
Fabrication and testing of satellites	1	4	8	
Farication and testing of satellite launchers				
Fabrication and testing of sounding rockets	2	4	12	
Construction of launch facilities				
Development of telemetry equipment		2	4	
Development of in-flight data acquisition and analysis facilities		2	1	
Development of inexpensive data collection platforms				
Development and launching of equatorial-orbiting satellites		1	2	
Development and launching of small satellites	1	3	6	
Development of scientific satellites		3	6	
Participation in joint space science experiments		2		

Total number of professional staff: 100   Total number of administrative staff: 20

Field of Activity Being Pursued

Field of activity	Description of major ongoing activities	Description of interaction with other developing contries and/or efforts towards addressing their needs
Fabrication and testing of satellites	Fabrication and testing of small scientific satellites (Badr series)	.
Fabrication and testing of satellite launchers		
Fabrication and testing of sounding rockets	Fabrication and testing of sounding rockets for upper atmosphere studies	

P

311

Pakistan

Field of activity	Description of major ongoing activities	Description of interaction with other developing contries and/or efforts towards addressing their needs
Construction of launch facilities		
Development of telemetry equipment	In support of sounding rockets for upper atmosphere studies	
Development of in-flight data acquisition and analysis facilities	Development of onboard computers for acquisition of telemetry for small satellites	
Development of inexpensive data collection platforms		
Development and launching of equatorial-orbiting satellites	A feasibility study for communication satellite system has been conducted. System is being implemented in the private sector.	
Development and launching of small satellites	Small multi-mission satellite (SMMS) project, with various payloads for remote sensing, GPS, communications, space environment, etc.	The SMMS project is being developed as a joint programme of several Asian and Pacific countries
Development of scientific satellites	A 50-kg LEO satellite with various scientific and experimental payloads	
Participation in joint space science experiments		

7.2.8 Republic of Korea

Satellite Technology Research Centre

The Satellite Technology Research Centre of the Korea Advanced Institute of Science and Technology is a university-associated organization mandated to train manpower in the space industry, undertake R and D in space science, remote sensing and small satellite technology and serve as the ESCAP national focal point for space technology applications.

Professional Personnel

Primary field of work	Number of staff with the following qualifications			Number of staff with intensive training (>2 consecutive months) during last 5 years
	Ph.D.	MSc	BSc	
Fabrication and testing of satellites				
Fabrication and testing of satellite launchers				
Fabrication and testing of sounding rockets				
Construction of launch facilities				
Development of telemetry equipment				
Development of in-flight data acquisition and analysis facilities				
Development of inexpensive data collection platforms				
Development and launching of equatorial-orbiting satellites				
Development and launching of small satellites	13	17	9	
Development of scientific satellites				
Participation in joint space science experiments				

Total number of professional staff: 39   Total number of administrative staff: 5

Field of Activity Being Pursued

Field of activity	Description of major ongoing activities	Description of interaction with other developing contries and/or efforts towards addressing their needs
Fabrication and testing of satellites	Development of KITSAT-3 satellites	
Fabrication and testing of satellite launchers		
Fabrication and testing of sounding rockets		
Construction of launch facilities		
Development of telemetry equipment		

R

Field of activity	Description of major ongoing activities	Description of interaction with other developing contries and/or efforts towards addressing their needs
Development of in-flight data acquisition and analysis facilities		
Development of inexpensive data collection platforms		
Development and launching of equatorial-orbiting satellites		
Development and launching of small satellites	Development of KITSAT-3 satellites	
Development of scientific satellites		
Participation in joint space science experiments		

Training Activities (1992-1996)

| Course title and description | Training undergone by organization staff | | |
	Given by (name of institution/agency)	Duration/ frequency	Number of staff taking course
System engineering	Lockheed Missile and Space Company	6 weeks	4
	Training provided for external users		
On-the-job training during KITSAT-2 project	Samsung Electronics	1 year	3
On-the-job training during KITSAT-2 project	Hyundai Electronics	1.5 years	7

Documents Published since 1990

Title	Year of publication	Language of publication	Keywords	Frequency
SaTReC Newsletter	1991-present	Korean		

Korea Aerospace Research Institute

The Korea Aerospace Research Institute (KARI) is a government-supported research institute mandated to conduct R and D on satellite, aircraft, scientific sounding rocket, and quality assurance; provide technical support for aerospace-related firms; and assist national policy in the aerospace field.

Professional Personnel

Primary field of work	Number of staff with the following qualifications			Number of staff with intensive training (>2 consecutive months) during last 5 years
	Ph.D.	MSc	BSc	
Fabrication and testing of satellites	3	12		5 (on-the-job training relevant to Koreasat project)
Fabrication and testing of satellite launchers				
Fabrication and testing of sounding rockets	10	10		
Construction of launch facilities				
Development of telemetry equipment	3	7		
Development of in-flight data acquisition and analysis facilities				
Development of inexpensive data collection platforms				
Development and launching of equatorial-orbiting satellites				
Development and launching of small satellites	28	40		
Development of scientific satellites				
Participation in joint space science experiments				

Total number of professional staff: 175   Total number of administrative staff: 62

Field of Activities Being Pursued

Field of activity	Description of major ongoing activities	Description of interaction with other developing contries and/or efforts towards addressing their needs
Fabrication and testing of satellites		
Fabrication and testing of satellite launchers		
Fabrication and testing of sounding rockets	Scientific sounding rocket programme (II) is continuing	
Construction of launch facilities		
Development of telemetry equipment	Korea Multi-purpose Satellite (KOMPSAT) ground station programme is under way	
Development of in-flight data acquisition and analysis facilities		
Development of inexpensive data collection platforms		
Development and launching of equatorial-orbiting satellites		
Development and launching of small satellites	KOMPSAT is continuing	KOMPSAT is being developed in cooperation with TRW in the United States
Development of scientific satellites		
Participation in joint space science experiments		

R

Training Activities (1992-1996)

Course title and description	Training undergone by organization staff		
	Given by (name of institution/agency)	Duration/ frequency	Number of staff taking course
Training programme on KOMPSAT project (structures, attitude control, assembly, launching and orbit, EMI and EMC test for KOMPSAT)	KARI (internal)	7 days	60
Training programme on design and testing of satellites	KARI (internal)	3 days, annual	60

Documents Published since 1990

Title	Year of publication	Language of publication	Keywords	Frequency
System design and development of the Korea multi-purpose satellite (I)	1995	Korean	Programme management of KOMPSAT, mission design and requirement analysis	
System design and development of the Korea multi-purpose satellite bus (I)	1995	Korean	System design of satellite bus and subsystems	
Assembly/integration/test facilities and technology development for the Korea multi-purpose satellite (I)	1995	Korean	Establishment of assembly, integration and test facilities for KOMPSAT	
Study on bus subsystems and fundamental technology development for communication satellites	1995	Korean	In-orbit operation and operation system	
System design and development of the Korea multi-purpose satellite (II)	1996	Korean	Mission design and requirement analysis, system engineering and its management	
System design and development of the Korea multi-purpose satellite bus (II)	1996	Korean	Design of satellite bus system and subsystems	
Assembly/integration/test facilities and technology development for the Korea multi-purpose satellite (II)	1996	Korean	Technology analysis of development of KOMPSAT test bed/EGSE and mechanical ground support equipment	
KOMPSAT ground station (I): installation and operation of data receiving facilities	1996	Korean	Requirement analysis and preliminary design for ground station	

7.2.9 Singapore

Nanyang Technological University (Satellite Engineering Research Programme)

The Nanyang Technological University is one of two universities providing tertiary education in Singapore. Details of its activities in this field may be obtained by accessing their Web site at <http://www.ntu.ac.sg>.

Professional Personnel

Primary field of work	Number of staff with the following qualifications			Number of staff with intensive training (>2 consecutive months) during last 5 years
	Ph.D.	MSc	BSc	
Fabrication and testing of satellites	17	1	4	3
Fabrication and testing of satellite launchers				
Fabrication and testing of sounding rockets				
Construction of launch facilities				
Development of telemetry equipment	17	1	4	3
Development of in-flight data acquisition and analysis facilities				
Development of inexpensive data collection platforms				
Development and launching of equatorial-orbiting satellites	17	1	4	3
Development and launching of small satellites	17	1	4	3
Development of scientific satellites				
Participation in joint space science experiments				

Total number of professional staff: 22   Total number of administrative staff: 2

Field of Activity Being Pursued

Field of activity	Description of major ongoing activities	Description of interaction with other developing contries and/or efforts towards addressing their needs
Fabrication and testing of satellites	L-band/S-band payload for UoSat-12 mini-satellite	The transponder will be available for access by developing countries
Fabrication and testing of satellite launchers		
Fabrication and testing of sounding rockets		
Construction of launch facilities		
Development of telemetry equipment	TT and C facility for LEO satellites	Small TT and C ground station being developed

S

Field of activity	Description of major ongoing activities	Description of interaction with other developing contries and/or efforts towards addressing their needs
Development of in-flight data acquisition and analysis facilities		
Development of inexpensive data collection platforms		
Development and launching of equatorial-orbiting satellites	Technical visibility study for a constellation of 6-8 satellites	Coverage involves all developing countries between 23°N and 23°S
Development and launching of small satellites	Future mini- and micro-satellite being planned	Common identical payloads can be available for regional satellites
Development of scientific satellites		
Participation in joint space science experiments		

Training Activities (1992-1996)

Course title and description	Training undergone by organization staff		
	Given by (name of institution/agency)	Duration/ frequency	Number of staff taking course
In-depth training in satellite communications	Singapore Telecom	66 hours	30, 25
Training in satellite communications	Ministry of Defence	12 hours	30

Documents Published since 1990

Title	Year of publication	Language of publication	Keywords	Frequency
UoSat-12 and the Merlion payload	1997	English	Mini-satellite, satellite transponders	
LEO/MEO equatorial orbit	1997	English	Small satellites, equatorial orbit	
Satellite activities at NTU	1996	English	Satellite projects	
LEO satellite research programme at NTU – the Merlion project	1996	English	Low earth orbit, MERLION project	
Investigation of the equatorial LEO orbit for small satellites	1996	English	Small satellites, equatorial orbit	
Exploration of equatorial LEO orbit for communications and other applications	1996	English	Small satellite applications, equatorial orbit, equatorial services	

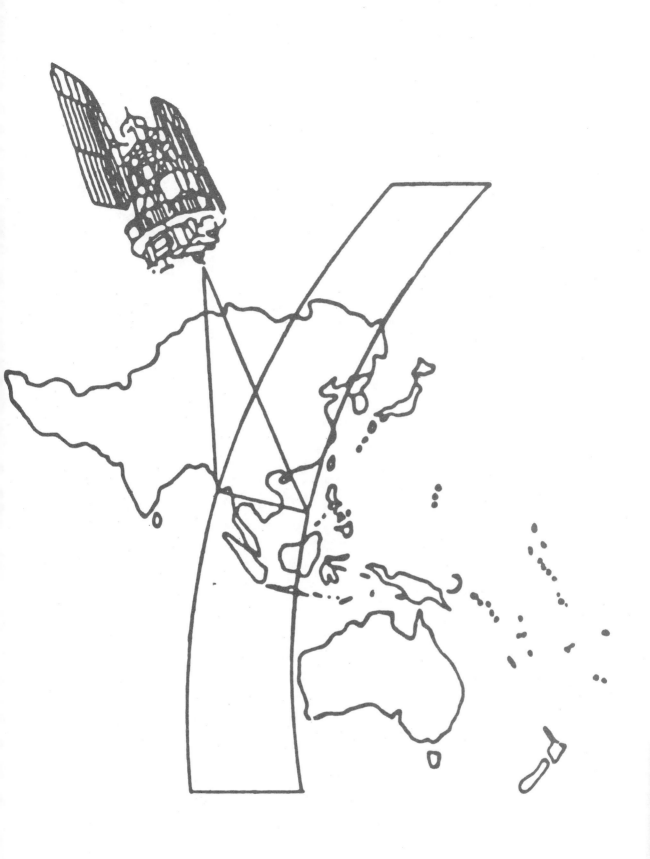

Annex I

BIBLIOGRAPHY

APT Yearbook, 1996, the Asia-Pacific Telecommunity, Bangkok.

Asiaweek (weekly magazine), various issues in 1996-1997.

Australian Space Research: 1994-1996, report to COSPAR, Australian National Committee for Space Science, Australian Academy of Science, 1996.

Bangkok Post (daily newspaper), various issues in 1996-1997.

Proceedings of the *Ministerial Conference on Space Applications for Development in Asia and the Pacific,* Beijing, 19-24 September 1994 (ST/ESCAP/1451).

CSIRO satellite data acquisition and utilization programme, annual report, 1994-1995 (final report), COSSA, Canberra, 1996.

Directory of Remote Sensing Education and Training in the ESCAP Region, 1990 (ST/ESCAP/1014).

EOSAT News. Vol. 11, No. 1, Spring/Summer 1996. Lanham, Maryland.

Inaugural report of Space Science Studies Division, Prime Minister's Department, Malaysia, 1994-1995.

Interface, bulletin from NRSA Data Centre, India, January-March 1996.

News from Prospace, No. 38, December 1995, Paris.

Pilot Scale Feasibility Study on An Earth Space Information Network for Asia and the Pacific, 1996 (ST/ESCAP/1712).

Proceedings of the First Meeting of the Regional Working Groups on Space Applications for Sustainable Development (ST/ESCAP/1728).

Proceedings of the Symposium on Space Technology and Applications in Sustainable Development, Beijing, 19-21 September 1994 (ST/ESCAP/1499).

Proceedings of the Workshop on Earth Science Information, organized by the Earth Science and Technology Organization, Colombo, 4-8 November 1996.

Regional Remote Sensing Inventory (Supplement 1990) (ST/ESCAP/1004).

Remote sensing: information from the Swedish Space Corporation, November 1996.

Report of the Annual Session of the Intergovernmental Consultative Committee on the ESCAP Regional Space Applications Programme for Sustainable Development in Asia and the Pacific (RESAP) and Proceedings of the Meeting of the Directors of the National Remote Sensing Centres/Programmes in the ESCAP Region, Dhaka, 10-16 June 1995 (ST/ESCAP/1559).

Report of the Eighth Session of the Intergovernmental Consultative Committee on the ESCAP/UNDP Regional Remote Sensing Programme and Proceedings of the Meeting of the Directors of the National Remote Sensing Centres/Programmes in the ESCAP Region, Hyderabad, India, 9-15 September 1991 (ST/ESCAP/1057).

Report of the Ninth Session of the Intergovernmental Consultative Committee on the ESCAP Regional Remote Sensing Programme and Proceedings of the Meeting of the Directors of the National Remote Sensing Centres/Programmes in the ESCAP Region, Islamabad, 8-12 May 1991 (ST/ESCAP/1291).

Report of the Second Session of the Intergovernmental Consultative Committee on the Regional Space Applications Programme for Sustainable Development; Report of the Meeting of the Regional Working Group on Remote Sensing, Geographic Information Systems and Satellite-based Positioning; and Summary Record of the Meeting of the Inter-agency Subcommittee on Space Applications for Sustainable Development in Asia and the Pacific, Kuala Lumpur, 3-8 June 1996 (ST/ESCAP/1689).

Report of the Seventh Session of the Intergovernmental Consultative Committee on the ESCAP/UNDP Regional Remote Sensing Programme and Proceedings of the Meeting of the Directors of the National Remote Sensing Centres/Programmes in the ESCAP Region, Kuala Lumpur, 25-30 June 1990 (ST/ESCAP/901).

Report of the Tenth Session of the Intergovernmental Consultative Committee on the Regional Remote Sensing Programme and Proceedings of the Meeting of the Directors of the National Remote Sensing Centres/Programmes in the ESCAP Region, Tehran, 22-26 May 1994 (ST/ESCAP/1414).

Report of the Working Group Meeting of the Regional Information Service and Education Networks of the Regional Remote Sensing Programme, Bangkok, 11-17 December 1990 (ST/ESCAP/973).

Report of the Working Group Meeting of the Regional Information Service and Education Networks of the Regional Space Applications Programme for Sustainable Development, Nakhon Ratchasima, Thailand, 20-23 November 1995 (ST/ESCAP/1619).

Russia's Military Space Forces: *Aerospace Journal.*

Space Policy, Vol. 12, No. 1, Oxford, United Kingdom, February 1996.

Space Research in Pakistan 1994 and 1995, national report to 31st COSPAR Scientific Assembly, Birmingham, United Kingdom, 14-21 July 1996; Pakistan Space and Upper Atmosphere Research Commission (SUPARCO), Karachi.

Space Science Activities in China: 1992-1993, national report, March 1994, Chinese National Committee for COSPAR, Beijing.

Space Technology and Applications for Sustainable Development in Asia and the Pacific: A Compendium, 1994 (ST/ESCAP/1420).

Space Technology Applications Newsletter (ESCAP), January, April, July and October 1996; January 1997.

STA, Its Roles and Activities: 1994-1995, Science and Technology Agency, Tokyo.

State of the Environment in Asia and the Pacific, 1990 (ST/ESCAP/917).

State of the Environment in Asia and the Pacific, 1995 (ST/ESCAP/1585).

World Almanac and Book of Facts 1995, The World Almanac, Mahwah, NJ, United States.

Annex II

LIST OF PERSONS WHO FILLED OUT
THE QUESTIONNAIRES

A. **Information on national remote sensing, GIS and satellite-based positioning activities was provided by the following persons:**

1. **Azerbaijan**

Mr Rustam B. Rustamov
Deputy Director General
Azerbaijan National Space Agency
159, Azadlyg Avenue, Baku 370106
Tel: (994-12) 62-93-87
Fax: (994-12) 62-17-38

2. **China**

Mr Ni Jian-Ping
Chief of Science and Technology Office
Changsha Branch, National Remote Sensing Centre
40, Xia Ma Yun Wang, Nan Oa Road, Changsha 41007
Tel: 86-073-151619
Fax: 86-073-5161590

Mr Li Qingquan
Director of Research Affairs Office
Wuhan Technical University of Surveying and Mapping
39 Lusyu Road, Wuhan 430070
Tel: 86-027-7881610
Fax: 86-073-5161590

Mr Zhao Xian Wen
Head of Department
Department of Remote Sensing of Forestry
Chinese Academy of Sciences, Beijing 100091
Fax: 86-10-62584972
E-mail: zhaoxul@info.forestry.ac.cn

3. **Fiji**

Mr Kemueli Masikerei
Principal Technical Officer
Ministry of Lands, Mining and Energy
P.O. Box 2222, Government Buildings, Suva
Tel: 314399
Fax: 305029
E-mail: ldanselm@itc.gov.fj

4. **Indonesia**

Mr Bambang Tejasukmana
Head, Remote Sensing Applications Centre
National Institute of Aeronautics and Space
Jl. Lapan, No. 70, Jakarta 13710
Tel: 62-21-8717717
Fax: 62-21-8717715

5. **Islamic Republic of Iran**

Mr Ahmad Mohammadpour
Deputy Director General in Application and GIS Affairs
Iranian Remote Sensing Centre
22, 14th Street, Saadat Abad Avenue
P.O. Box 11365/6713, Tehran
Tel: 2063208
Fax: 2064474

6.	**Japan**	Mr Sohsuke Gotoh

Mr Sohsuke Gotoh
Special Assistant to the President
National Space Development Agency of Japan
2-4-1, Hamamatsu-cho, Minato-ku, Tokyo 105-60
Tel: 81-3-5401-8704
Fax: 81-3-5401-8702
E-mail: gotoh@nsaeoc.eoc.nasda.go.jp

Mr Koji Takimoto
Deputy Director of Space Space Industry Division
Ministry of International Trade and Industry
1-3-1 Kasumigaseki, Chiyoda-ku, Tokyo 100, Japan
Tel: 81-3-3501-0973
Fax: 81-3-3501-6723
E-mail: tkaa9392@miti.go.jp

7. **Mongolia**

Mr M. Badarch
Director
National Remote Sensing Centre
Khudaldaany Gudamj-5, Ulaanbaatar 11
Tel: 976-1-328151
Fax: 976-1-323189
E-mail: agenda21@magicnet.mn

8. **Myanmar**

U Htay Aung
Deputy Director General
Department of Meteorology and Hydrology
Kaba-Aye Pagoda Road, Yangon
Tel: 95-1-660823
Fax: 95-1-665944

9. **Nepal**

Mr Rajendra B. Joshi
Executive Director
Forest Research and Survey Centre
P.O. Box 3339, Babar Mahal, Kathmandu
Tel: 977-1-220493
Fax: 977-1-220159
E-mail: fris@mos.om.np

10. **Pakistan**

Mr M. Ishaq Mirza
Member, Space Research
Pakistan Space and Upper Atmosphere Research Commission
P.O. Box 8402, Sector 28, Gulzar-e-Hijri, Off University Road, Karachi-75270
Tel: 92-21-8141311, 4968667
Fax: 92-21-4960553, 4965638
E-mail: suparco@biruni.erum.com.pk

11. **Philippines**

Ms Virginia Sicat-Alegre
Chief, Information Services Division
National Mapping and Resource Information Authority
NAMRIA Building, Lawton Avenue, Fort Andres Bonifacio, Makati
Tel: 63-2-8104831
Fax: 63-2-8102891
E-mail: namria@sun1.dost.gov.ph

Mr Elson Q. Aca
Manager, Project Development and GIS Management Services
Geodata Systems Technologies, Inc.
19/F Strata 100 Building, Emerald Avenue, Ortigas Centre, Pasig City 1605
Tel: 63-2-632-4947 to 49
Fax: 63-2-633-6873
E-mail: geodata@mozcom.com

Mr Reuben Campos
Research Associate
Marine Science Institute
University of the Philippines
Diliman, Quezon City, Metro Manila
Tel: 63-2-922-3959
Fax: 63-2-924-7678, 924-3735
E-mail: camposb@msi.upd.edu.ph

Mr Oliver Coroza
Project Manager
Cybersoft Integrated Geoinformatics, Inc.
6th Floor, Triumph Building, Quezon Avenue, Quezon City
Tel: 63-2-929-5092
Fax: 63-2-929-5011

Ms Alma V. Nepomuceno
Engineer III
National Water Resources Board
8th Floor, NIA Building, EDSA, Quezon City 1100
Tel: 63-2-920-2681
Fax: 63-2-920-2641

Mr Randy John N. Vinluan
Research Associate
Training Centre for Applied Geodesy and Photogrammetry
University of the Philippines
Diliman, Quezon City, Metro Manila
Tel: 63-2-434-3633
Fax: 63-2-922-4714, 928-3144
E-mail: rjnv@engg.upd.edu.ph

12.	**Republic of Korea**	Mr Tae-Yong Kwon Research Scientist Systems Engineering Research Institute P.O. Box 1, Yoosung, Taejon 305-333 Tel: 82-42-8691963 Fax: 82-42-8691460 E-mail: tykwon@seri.ie.kr
13.	**Singapore**	Mr Kwoh Leong Keong Assistant Director Centre for Remote Imaging, Sensing and Processing National University of Singapore Lower Kent Ridge, Singapore 119260 Tel: 65-7727838 Fax: 65-7757717 E-mail: crsklk@leonis.nus.sg
14.	**Sri Lanka**	Mr H. Manthrithilake Director, Environment and Forest Conservation Division Mahaweli Authority Dam Site, Polgolla Tel: 08-24950 Fax: 08-234950 E-mail: efcdmasl@slt.lk

15.	**Thailand**	Mr Rasamee Sunannerakamtorn
		Senior Research Scientist
		Thailand Remote Sensing Centre
		196 Phaholyothin Road, Chatuchak, Bangkok 10900
		Tel: 66-2-4542810
		Fax: 66-2-5613035
		E-mail: application@fe-nrct.go.th

16.	**Viet Nam**	Mr Nguyen Van Hieu
		President
		National Centre for Science and Technology (NCST) of Viet Nam
		P.O. Box 607, Bo Ho, Hanoi
		Tel: 84-4-8361779
		Fax: 84-4-8352483

B. Information on national satellite meteorology and disaster monitoring activities was provided by the following persons:

1.	**Azerbaijan**	Mr Rustam B. Rustamov
		Deputy Director General
		Azerbaijan National Space Agency
		159, Azadlyg Avenue, Baku 370106
		Tel: 994-12-62-93-87
		Fax: 994-12-62-17-38

2.	**Bangladesh**	Mr A.M. Choudhury
		Chief Scientific Officer and Head, Research
		Bangladesh Space Research and Remote Sensing Organization
		Mohakash Biggyan Bhaban, Agargaon, Sher-E-Bangla Nagar, Dhaka 1207
		Tel: 880-2-323978
		Fax: 880-2-813080
		E-mail: drsdc@bas.bdmail.net

3.	**China**	Mr Xu Jianmin
		National Satellite Meteorological Centre
		46, Baishiqiaolu, Haidian, 100081 Beijing
		Tel: 62173894
		Fax: 62172724

Mr Ni Jian-Ping
Chief of Science and Technology Office
Changsha Branch, National Remote Sensing Centre
40, Xia Ma Yun Wang, Nan Oa Road, Changsha 41007
Tel: 86-073-151619
Fax: 86-073-5161590

Mr Zhang Jianzhong
Vice Chief of Project Office
Institute of Remote Sensing Applications
P.O. Box 9718, Beijing 100101
Tel: 64919458
Fax: 64915045

Dr Liu Liang Ming
National Laboratory for Information Engineering in Surveying,
Mapping and Remote Sensing
Wuhan Technical University of Surveying and Mapping
39 Luo-yu Road, Wuhan 430070
Tel: 86-027-7881292
Fax: 86-027-7814185

4.	Indonesia	Mr Agus Hidayat Senior Scientist National Institute of Aeronautics and Space Jl. Lapan, No. 70, Jakarta 13710 Tel: 62-21-8710065 Fax: 62-21-8717715

Mr M. Nazamuddin
Chief of Processing Administration
Meteorological and Geophysical Agency
Jl. Angkasa I/No. 2, Jakarta Pusat, Jakarta 13710
Tel: 62-21-3156156
Fax: 62-21-3107788
E-mail: bmg@cbn.net.id

5. Islamic
    Republic
    of Iran

Ms Mehrnaz Bijanzadeh
Head of Satellite Meteorology Centre
Islamic Republic of Iran Meteorological Organization
P.O. Box No. 13185-461, Tehran
Tel: 98-21-6469047
Fax: 98-21-6469044

6. Japan

Mr Tetsu Hiraki
Head of the Office of Meteorological Satellite Planning
Japan Meteorological Agency
1-3-4 Otemachi, Chiyoda-ku, Tokyo 100
Tel: 81-3-3201-8677
Fax: 81-3-3217-1036
E-mail: cgms@jma.hq.kishou.go.jp

7. Malaysia

Mr Alui Bahari
Director
Malaysian Meteorological Service
Jln. Sutran, 46667 P. Jaya
Tel: 9569422
Fax: 7578046
E-mail: alui@kjc.gov.my

8. Mongolia

Mr M. Badarch
Director
National Remote Sensing Centre
Khudaldaany Gudamj-5, Ulaanbaatar 11
Tel: 976-1-328151
Fax: 976-1-323189
E-mail: agenda21@magicnet.mn

9. Nepal

Mr Kiran Shankar Yogacharya
Director-General
Department of Hydrology and Meteorology
P.O. Box 406, Babar Mahal, Kathmandu
Tel: 977-1-212151
Fax: 977-1-224648

10. Pakistan

Mr M. Ishaq Mirza
Member, Space Research
Pakistan Space and Upper Atmosphere Research Commission
P.O. Box 8402, Sector 28, Gulzar-e-Hijri, Off University Road, Karachi-75270
Tel: 92-21-8141311, 4968667
Fax: 92-21-4960553, 4965638
E-mail: suparco@biruni.erum.com.pk

11.	**Republic of Korea**	Mr Ae-Sook Suh Director, Remote Sensing Research Laboratory Korea Meteorological Research Institute 2, Wanyoung-Dong, Chongno-gu, Seoul 110-360 Tel: 797-7841 Fax: 3672-8083 E-mail: suh@tms.metri.re.kr
12.	**Singapore**	Mr Koo Hock Chong Deputy Director (Science) Meteorological Service Singapore P.O. Box 8, Singapore Changi Airport 918141 Tel: 65-5457196 Fax: 65-5457192 E-mail: kstan@cs.gov.sg
13.	**Sri Lanka**	Mr N.A. Amaradasa Deputy Director Department of Meteorology 383, Bauddhaloka Mawatha, Colombo 7 Tel: 692756 Fax: 691443
14.	**Viet Nam**	Mr Hoang Minh Hien Researcher in Satellite Meteorology National Centre for Hydrometorological Forecasting 4 Dang Thai Than Street Hanoi Tel: 84-4-8253343 Fax: 84-4-8260779

C. Information on national satellite communications and distance education activities was provided by the following persons:

1.	**Azerbaijan**	Mr Rustam B. Rustamov Deputy Director General Azerbaijan National Space Agency 159, Azadlyg Avenue, Baku 370106 Tel: 994-12-62-93-87 Fax: 994-12-62-17-38
2.	**China**	Mr Cui Xiaohui Engineer Wuhan Technical University of Surveying and Mapping 39 Lusyu Road, Wuhan 430070, China Tel: 86-27-7861414 Fax: 073-5161590 E-mail: xhcui@dns.wtusm.edu.cn
		Mr Ni Jian-Ping Chief of Science and Technology Office Changsha Branch, National Remote Sensing Centre 40, Xia Ma Yun Wang, Nan Oa Road, Changsha 41007 Tel: 86-073-151619 Fax: 86-073-5161590

3.	**Indonesia**	Mr Denny Setiawan
		Frequency Planning Section
		Ministry of Posts, Tourism and Telecommunication
		Jl. Medan Merdeka Barat, No. 16-19, Jakarta 10110
		Tel: 62-21-383-8367
		Fax: 62-21-386-7500
		E-mail: renfrek@indosat.net.id

4. **Islamic Republic or Iran** Mr Abdol Mohammad Darab
President
Iran Telecommunication Research Centre
P.O. Box 14155-3961
Tehran
Tel: 78-21-841-717
Fax: 78-21-800-9930
E-mail: itrc-lib@rose.ipm.ac.ir

5. **Japan** Mr Takamasa Suzuki
Assistant Senior Director
Telecommunication Advancement Organization of Japan
2-31-19 Shiba, Minato-ku, Tokyo, Japan
Tel: 81-3-3769-6820
Fax: 81-3-5491-7584

6. **Mongolia** Mr Sanjaa Ganbaatar
President
Mongolian Telecommunication Company
P.O. Box 1166, Sq. Sukhbaatar-9, Ulaanbaatar-210611
Tel: 976-1-324856
Fax: 976-1-325412
E-mail: ganbatar@magicnet.mn

7. **Nepal** Mr B.R. Kanel
Director, Telecom Training Centre
Nepal Telecommunication Corporation
Babar Mahal, Kathmandu
Tel: 977-1-221809
Fax: 977-1-223360

8. **Pakistan** Mr S. Akram Hussain Abidi
Member, Space Electronics
Pakistan Space and Upper Atmosphere Research Commission
P.O. Box 8402, Sector 28, Gulzar-e-Hijri, Off University Road, Karachi-75270
Tel: 92-21-4964174
Fax: 92-21-4960553
E-mail: suparco@biruni.erum.com.pk

9. **Singapore** Professor E.S. Seumahu
Principal Investigator
School of Electrical and Electronic Engineering
Nanyang Technological University
Nanyang Avenue, 639798
Tel: 65-7995436
Fax: 65-7920415
E-mail: eseumahu@ntu.edu.sg

10.	**Sri Lanka**	H.S. Padmasiri de Alwis
		Deputy Director
		Arthur C. Clarke Centre for Modern Technologies
		Katubedda, Moratuwa
		Tel: 94-1-647-678
		Fax: 94-1-647-462
		E-mail: accmt@mail.ac.lk

11. **Thailand**

Mr Sanong Chinnanon
Director
Thaicom Distance Education Centre
c/o Centre for Educational Technology
Sri Ayudhaya Road, Rajthewee, Bangkok 10400
Tel: 66-2-2477064
Fax: 66-2-3477064

International Services Division
Post and Telegraph Department
87, Phaholyothin Soi 8, Phayathai, Bangkok 10400
Tel: 66-2-984-8028
Fax: 66-2-984-8030

Telephone Organization of Thailand
26/16-17 Oragan Building, Chidlom Road, Pratumwon, Bangkok 10300

Mr Traichuk Prommakoun
Director of Engineering Division
Communication Authority of Thailand
99, Chaeng Wathana, Don Muang, Bangkok 10002
Tel: 66-2-506-4111
Fax: 66-2-574-4794

12. **Viet Nam**

Nguyen Thanh Phuc
Manager
Department General of Posts and Telecommunications
18 Nguyen Du Street, Hanoi
Tel: 84-4-822-8383
Fax: 84-4-822-6590
E-mail: ch.vienkhon@bdvn.vnd.net

D. Information on national space science and technology activities was provided by the following persons:

1. **Azerbaijan**

Mr Rustam B. Rustamov
Deputy Director General
Azerbaijan National Space Agency
159, Azadlyg Avenue, Baku 370106
Tel: 994-12-62-93-87
Fax: 994-12-62-17-38

2. **Bangladesh**

Mr Nazmul Hoque
Chief Scientific Officer
Bangladesh Space Research and Remote Sensing Organization
Mohakash Biggyan Bhaban, Agargaon, Sher-E-Bangla Nagar, Dhaka 1207
Tel: 880-2-323942
Fax: 880-2-813080

3.	**Indonesia**	Soewarto Hardhienata Head, Space Communication Technology Division National Institute of Aeronautics and Space Jl. Lapan, No. 70, Jakarta 13710 Tel: 62-251-62166 Fax: 62-251-623010
4.	**Japan**	Mr Kazuo Anazawa Head, International Affairs Division Administration Department Institute of Space and Astronautical Science 3-1-1 Yoshinodai, Sagamihara, Kanagawa 229 Tel: 81-4-2751-5022 Fax: 81-4-2752-4363 E-mail: kokusaia@adm.isas.ac.jp
5.	**Mongolia**	Mr M. Badarch Director National Remote Sensing Centre Khudaldaany Gudamj-5, Ulaanbaatar 11 Tel: 976-1-328151 Fax: 976-1-323189 E-mail: agenda21@magicnet.mn
6.	**Pakistan**	Mr M. Nasim Shah Secretary Pakistan Space and Upper Atmosphere Research Commission P.O. Box 8402, Sector 28, Gulzar-e-Hijri, Off University Road, Karachi-75270 Tel: 92-21-4965059 Fax: 92-21-4960553 E-mail: suparco@biruni.erum.com.pk
7.	**Republic of Korea**	Mr Taejung Kim Associate Senior Researcher Satellite Technology Research Centre (SaTReC), KAIST 373-1, Kusung-Dong, Yusung-Gu, Taejon Tel: 82-42-8698629 Fax: 82-42-8610064 E-mail: tkim@krsc.kaist.ac.kr
8.	**Singapore**	Professor E.S. Seumahu Principal Investigator School of Electrical and Electronic Engineering Nanyang Technological University Nanyang Avenue, 639798 Tel: 65-7995436 Fax: 65-7920415 E-mail: eseumahu@ntu.edu.sg

A	Australia
	Azerbaijan
B	Bangladesh
C	China
F	Fiji
I	India
	Indonesia
	Islamic Republic of Iran
J	Japan
M	Malaysia
	Mongolia
	Myanmar
N	Nepal
	New Zealand
P	Pakistan
	Philippines
R	Republic of Korea
	Russian Federation
S	Singapore
	Sri Lanka
T	Thailand
V	Vanuatu
	Viet Nam